CAMBRIDGE LIBRARY COLLECTION

Books of enduring scholarly value

Life Sciences

Until the nineteenth century, the various subjects now known as the life sciences were regarded either as arcane studies which had little impact on ordinary daily life, or as a genteel hobby for the leisured classes. The increasing academic rigour and systematisation brought to the study of botany, zoology and other disciplines, and their adoption in university curricula, are reflected in the books reissued in this series.

Catalogue of Plants Cultivated in the Garden of John Gerard, in the Years 1596-1599

For twenty years, the herbalist John Gerard (1545–1612) served as superintendent of the gardens of Elizabeth I's minister, Lord Burghley. The 1596 edition of Gerard's *Catalogus* is probably the first complete catalogue of any one garden, public or private, ever published. Describing his own garden, the list includes frankincense, saffron, an almond tree and even tulips, then exotic and notoriously costly. Probably intended originally only for the interest of Gerard's friends, and containing numerous errors, it progressed in 1599 into a new, improved edition for a much wider readership. In this book, first published privately in 1876, the botanist Benjamin Daydon Jackson (1846–1927) reproduces both editions, preserving the original errors and adding a memoir of the author that demonstrates the depth of his own research. With the modern names of the plants printed beside their earlier counterparts, Jackson's text is a fascinating resource for historical botanists and taxonomists.

Catalogue of Plants Cultivated in the Garden of John Gerard, in the Years 1596-1599

Edited by Benjamin Daydon Jackson

CAMBRIDGE
UNIVERSITY PRESS

CAMBRIDGE UNIVERSITY PRESS

Cambridge, New York, Melbourne, Madrid, Cape Town,
Singapore, São Paolo, Delhi, Tokyo, Mexico City

Published in the United States of America by Cambridge University Press, New York

www.cambridge.org
Information on this title: www.cambridge.org/9781108037150

This edition first published 1876
This digitally printed version 2011

ISBN 978-1-108-03715-0 Paperback

A

CATALOGUE OF PLANTS

CULTIVATED

IN THE GARDEN OF

JOHN GERARD,

In the years 1596—1599.

EDITED

WITH NOTES, REFERENCES TO GERARD'S *HERBALL*, THE

ADDITION OF MODERN NAMES,

AND

A LIFE OF THE AUTHOR,

BY

BENJAMIN DAYDON JACKSON, F.L.S.

Privately printed.

LONDON.

1876.

LONDON:
Pewtress and Co., Printers, 15, Gt. Queen Street,
Lincoln's Inn Fields, W.C.

CONTENTS.

INTRODUCTION.

In issuing this edition of an exceedingly scarce and interesting work, a short account of the book itself, and of the Editor's additions, will be expected.

My aim has been to give an exact copy of the first edition of Gerard's Catalogue of his Garden, line for line, and letter for letter, carefully retaining the printer's errors, but not attempting an absolute facsimile. The various borders, head and tailpieces, and the type, are imitated as nearly as could be done with modern appliances, but I have not copied turned letters, letters from a wrong fount, nor certain curious braces, which, no doubt intended to bracket together nearly allied species, were strangely misapplied. The original printing is very bad, in some places very black, in others as faint; Italic, and small capitals are frequently used in place of Roman; as the printer, Robinson, was living at that time in "Fewter lane," not very far from Gerard's abode, that was probably the reason for the work being given to him. Lord Burleigh's coat of arms, on verso of title, which I have omitted, was worked from the same block which was employed for the Herball in the following year.

Next in order will be found a reprint of the second edition of the same work, having in the original two vertical columns, the first in Italic type, containing the Latin names, the second, the English equivalent; I have followed this plan, so far as the style of type is concerned, but on a smaller scale; and for the sake of economising space, I have not repeated the leading name of each plant, when it occurs more than once, but merely give the initial, according to modern usage. Following the Gerardian names, will be found references to the Herball (1597), so far as I have been able to correlate the plants of the Catalogue with those of the larger work, no easy matter in many cases from the names failing to correspond; the references are to the correct numbers of the pages which are sometimes at variance with the printed numbers. A dash ——— following the English name, indicates that I am not able to quote the plant as occurring in the Herball; where Johnson criticises Gerard, gives a better figure, or supplies one wanting in the Author's edition, I have also quoted the edition of 1633. It will be seen that extracts from the Herball are frequently supplied, when likely to add interest to the account of the several plants. In a few other cases, I have referred to the other works, for which I must refer the reader to the list of books quoted. With very few exceptions, I have strictly adhered to the original order in which the items occur, an occasional transposition having been made for the sake of convenience, and a very few obvious misprints corrected.

The names given last are in Clarendon type, and are those in modern use. In reducing the old nomenclature to its modern equivalents, I have made free use of contemporaneous botany. Johnson's edition of Gerard's Herball (1633), Bauhin's Pinax (1623), and Ray's Historia Plantarum (1686-1704), were in constant requisition, whilst Aiton's Hortus Kewensis, ed. II. (1810-13) was of great service in checking the results. Many otherwise doubtful plants have been determined by help of the Sloane Herbarium, the possession of which, contributes to render the British Museum unrivalled for such researches; the only other place, in this country at all events, where similar facilities are obtainable, is Oxford, from the Sherardian Library and Herbarium in the Botanic Garden there. The vagueness noticeable in Gerard's works, has proved a constant source of annoyance and possible error in the task of determination; I have, however, succeeded in escaping some mistakes which my predecessors in this unfrequented field have committed. How far I have completely avoided error, I of course cannot tell, but I may honestly state, that I have spared neither time nor pains to render this work as accurate throughout as possible; those who have engaged in similar work to mine, will, I doubt not, look leniently upon the shortcomings of this work, to which I cannot blind myself. It has frequently been necessary, from the total incompatibility of the English and Latin names, to judge by probabilities, which denomination to follow.

The typographical execution of the second edition is far superior to that of the first, as will be seen on inspection even of this reprint; the original pagination I have indicated by figures placed in the margin. The only copy of the first edition, so far as I am aware, is that in the British Museum, (Press mark, C. 18. *b.*) formerly in the possession of Sir Hans Sloane; this is the copy described by Dryander in

Bibl. Banks, and followed by Pritzel, there is also a MS. copy in the Banksian collection, in the Botanical Department of the same institution. The second edition although far from common, is to be met with occasionally in private hands, as well as in public libraries. It is erroneously described by Pritzel in both editions of his Thesaurus, from the copy in the Bodleian Library, as a quarto; it is really a small folio. An example to which great interest is attached, is bound up with a copy of the Herball, now in the Botanical Department of the British Museum, and formerly belonging to James Petiver; this has suffered from the depredations of mice, and injudicious trimming of the gnawed edges to conceal that injury as much as possible. Petiver has written on the title page "Ex dono generosi D. R. Reynardson," and the epithet "cornicula æsopica Clusii" has been added to the author's name by some earlier owner. A few MS. notes are scattered through this copy, and the certificate has been defaced as mentioned on p. 55, and as I believe by L'Obel himself.

From its rarity, and careless printing, I should infer that the édition of 1596 was chiefly intended to circulate among the author's own immediate friends; but after the Herball appeared, the amended edition of 1599 would command a much larger circulation, and the copies would be more carefully preserved.

It may be worth remarking that the weather about the time when Gerard published his Catalogus and Herball, was most unpropitious; for a series of years, wet summers had raised the price of corn, and in 1596, wheat in London reached the famine price of £5 · 4 0 per quarter; this too when the purchasing power of money was fully six times its present value. This fact should be borne in mind, when noticing the Author's statements about his garden.

The memoir of the Author has been drawn up from all available sources, the Herball largely contributing; from contemporary writers a good deal has been obtained; the parish registers both of his native place and his residence in London, have been utilised, and some information has been gleaned from the Public Record Office, whilst the College of Arms, and the Registers of the Stationers' Company have been laid under contribution. The chief source of hitherto unpublished information relative to Gerard, has been gained from the Archives in the possession of the Barbers' Company. Had these last been in better order, and in worthier keeping, probably more might have been ascertained than I have been enabled to do. At the time of Gerard's connection with this Livery Company, its power was at its zenith; but as time passed on, it proved inadequate to fufil the requirements of the day, and like the great majority of similar corporations, now only exists for the gratification of its members.

The circumstances attending my search in the Barbers' Records are worthy of remark. In the first instance, I wrote to the Clerk of the Company, Mr. HENLEY GROSE SMITH, asking if he could give me any information about the subject of my work; after waiting for more than a week, without receiving any answer, I went to the Hall in Monkwell Street, and was thence referred by the housekeeper to the Clerk's office in the City. In my interview with him, he said that he was pressed for time just then, and that if he sent a clerk to the Hall purposely, he must charge his time to me, but that if I would consent to wait a fortnight, his clerk would be at Barbers' Hall, and I could " take advantage of his being there." Although the delay was irksome, I consented to wait, but shortly afterwards, Mr. SMITH, for his own convenience, wrote to postpone the date of my visit, to nearly a month subsequent to my interview with him. On the day appointed, I carefully looked over such of the books about the required date as the clerk could find, and quitted the place in less than three hours from the time of entering. The following day to my extreme surprise, I received a note from the Clerk, requesting me to remit one guinea "for clerk's time." I replied, pointing out how monstrous it was to charge me with an exorbitant amount, merely for his deputy's trouble in finding and placing before me certain books, for whilst I was at the Hall, the Clerk's clerk was engaged in work for the Company, apparently preparing notices to be sent to the members. For answer to this letter of mine, I had an offensively worded epistle, totally ignoring the questions I had put, and taking credit for making no charge for correspondence. In rejoinder I said, that the so-called correspondence consisted of these items: (a) *Not* answering my letter of enquiry, and thus compelling me to call at two places, (b) altering the date of my visit to suit himself, (c) sending in his claim, (d) his attempt to justify it. Although strongly urged to withstand this imposition, I felt it impossible to remain under the slightest obligation in this quarter, and therefore sent a cheque, enclosed in a brief note, expressing my sense of the want of right feeling displayed by twice charging for his clerk's services, first to the Company and then to me, who neither had, nor required, the undivided attention of his subordinate. I received no acknowledgement whatever of this letter.

There only remains the pleasing duty of acknowledging the kind assistance of those who have so materially contributed to the completion of my task. To the officers of the Botanical Department of the British Museum, Mr. WILLIAM CARRUTHERS, F.R S., Dr. HENRY TRIMEN, and Mr. JAMES BRITTEN, I must express my deep obligation for help afforded during the entire period of preparation; to Mr. John Gilbert Baker, of the Kew Herbarium, for assistance in several doubtful points; to the Rev. Foster Gray Blackburn, M.A., Rector of Nantwich, for obliging information as to the Registers of that place; to Mr. Joseph Greenhill, the Registrar of the Stationers' Company, for permission to inspect the Registers under his care; and lastly to Mr. Henry Wicks, of the Firm of Pewtress & Co., for the time and attention he has bestowed upon the printing.

B. DAYDON JACKSON.

30, Stockwell Road,

London, S.W.

June, 1876.

LIST OF
QUOTED BOOKS AND AUTHORITIES.

Barb. MS.	Records of the Barbers' Company.
Barb. List.	List of Masters and Wardens of the Barbers' Company.
Bauh. Pin.	Bauhin, Pinax, . . . Basiliæ, 1623.
Biog. Brit.	Biographia Britannica . . . London, 1747-66.
Coles.	Coles, W., Adam in Eden, . . . London, 1657.
Cordi Annot.	Cordus, V., Annotationes . . . Argent, 1561.
Fl. Middx.	Trimen and Dyer, Flora of Middlesex, London, 1868.
Ger.	The Herball, . . . by John Gerard . . . London, 1597.
Ger. Cat. ed. I.	Catalogus . . . London, 1596.
Ger. Cat. ed. II.	Ibid. ,, 1599.
Ger. em.	The Herball . . . enlarged and emended by Thomas Johnson, London, 1633.
Haller, Bibl. Bot.	Haller, A., Bibliographia Botanica . . . Tiguri, 1771.
Hentz. Itin.	Hentzner, Itinerarium, . . . Breslau, 1617.
Lansd. MSS.	Lansdowne MSS. British Museum.
Lob. Adv.	L'Obel and Pena, Adversaria nova, London, 1570.
Lob. Adv. Alt. pars.	L'Obel, Adversariorum altera pars, London, 1605.
Lob. Ill.	L'Obel, Illustrationes, London, 1655.
Lob. MS.	L'Obel, MS., notes in a copy of Gerard's Catalogus, Ed. II. v. p. vii.
Lob. Rond.	L'Obel, . . . Rondeletiana, [in Lob. Adv. Alt. pars.]
Lyte.	A niewe herball, . . . by Henry Lyte, . . . London, 1578.
Maitland	Maitland, W., History of London, . . . London, 1739.
Mill. Gard. Dict.	Miller, P., Gardener's Dictionary, Ed. VIII. London, 1768.
Noorthouck	Noorthouck, J., History of London, Westminster and Southwark . London, 1773.
Park. Par.	Parkinson, J., Paradisus terrestris . . . London, 1629.
R. Hist.	Ray, J., Historia plantarum . . . London, 1686-1704.
Rees, Cyclo.	Rees, A., Cyclopædia . . . London, 1819.
Salm.	Salmon, W., Botanologia . . . London, 1710-11.
Spreng. Hist.	Sprengel, C., Historia Rei. Herbariæ Amst, 1807-8.
Stow.	Stow, J., A survaye of London. . . . London, 1598.
Strype.	A survey of London and Westminster, by J. Stow, edited by J. Strype. London, 1720.

A

LIFE

OF

THE AUTHOR.

JOHN GERARD was born at Nantwich,[1] in the county of Cheshire, in 1545,[2] but owing to the loss of the baptismal registers of that place, from May in that year to 1572, we are not able to give any closer approximation to the precise date. He was descended from some younger branch of the Gerards of Ince, in Lancashire, as we learn from his own Coat of Arms,[3] which bears a crescent for difference, the crest, a lion's gamb, erased, inverted, holding a hawk's lure, with the motto *D'assenti buone*. There are no records at the College of Arms to shew his parentage. His name is most frequently spelled "Gerarde," but this mode arises, no doubt, from an engraver's error in the Title-page, for Gerard himself, and his friends, invariably spelled his name without the "e" final.

He went to school at Wisterson, or Willaston of the Ordnance Survey, two miles from Nantwich,[4] and probably there received all his scholastic education.[5] At an early age he was drawn to the study of medicine,[6] and travelled, possibly as ship's surgeon, on board some merchant vessel trading northwards, since he speaks of having been from "Narua vnto Moscouia, . . . the Sownde, beyonde Denmarke,"[7] and again, "Denmarke, Swenia, Poland, Liuonia, or Russia, or in any of those colde countries, where I haue travelled;"[8] he may possibly also have visited the Mediterranean.[9]

He must have settled in London before 1577, since he speaks of having superintended the gardens belonging to Lord Burleigh, in the Strand, and at Theobalds in Hertfordshire, for twenty years, which occupation took up the greatest part of his time.[10] His employer expended £10 weekly, to keep the poor employed in his gardens; he also says, "For my servants, I keep none to whom I pay not wages, and give liveries, which I know many do not."[11] L'Obel mentions Lord Burleigh's garden in London,[12] but as he is totally silent in his earlier works about Gerard, it affords additional proof that the latter had not then made his mark as a successful gardener. Hentzner, who visited Theobalds, Sept. 8th, 1598, the day Lord Burleigh was buried, has given a description of the garden at that place, at the time when Gerard was the superintendant.[13] There is no existing list of admissions to the Freedom and Livery of the Barber-Surgeons' Company, about this time, so that we cannot from this source ascertain the date of his first residing in London; at that period, no one could carry on the trade of "barbarie or chirurgerie" in the City, without being at least a Freeman of that Company; recalcitrant members of the craft being summarily committed to the Compter.

He made friends in his profession, for George Baker, "one of the chiefe chirurgions in ordinarie" to Queen Elizabeth, had a high opinion of Gerard's attainments, since he says,[14] "I protest vpon my conscience, I do not thinke for the knowledge of plants, that he is inferior to any: for I did once

1 Ger. 203. 2 Portrait in Ger. ante p. 1. 3 Ib. lefthand lower corner. 4 Ger. 1091. 5 Ib. 315.
6 Ger. cat. ed. II. præf. 7 Ger. 1177. 8 Ib. 1223. 9 Ib. 1171. 10 Ib. pref.
11 Biog. Brit. ii. 1267. 12 Lob. Adv. 422. 13 Hentz. Itin. 138. 14 Baker, in Ger.

see him tried with one of the best strangers that euer came into England, and was accounted in Parise the onely man, being recommended to me by that famous man M. Amb. Pareus, and he being here was desirous to go abroad with some of our herbarists, for the which I was the meane to bring them togither; and one whole day we spent therein, searching the most rarest simples: but when it came to the triall, my French man did not know one to his fower." This statement can only apply to Jean Robin, who, in 1597, was appointed keeper of the King's garden in Paris, on account of his success as a cultivator; that Gerard was on intimate terms with Robin, may be seen in the number of plants received from him, and acknowledged in the Herball.

Gerard was elected a member of the Court of Assistants of the Barber-Surgeons, June 19th, 1595,'[15] and in the following year, July 16th, 1596,[p] he was commissioned, with another, to seek a better place for a "fruit-ground" than that in "East Smithfielde or ffetterlane." At this time he had a house in Holborn,[16] then the most aristocratic portion of London; his garden was probably attached to his house, or may have been the identical plot mentioned above, in Fetterlane, as there were many gardens belonging to the wealthier citizens in that locality. The northern side of Holborn, which has been suggested as the probable site of Gerard's garden, is unlikely, since during the last few years of Elizabeth's reign, the spot now occupied by Ely Place and Hatton Garden, was a garden of forty acres, belonging to the Bishopric of Ely.[17]

The following interesting draft of a letter is in Gerard's autograph, and was probably drawn up previous to 1596:—

"After my most hartie commendacions, &c. As yt hath beene alwaies myne especiall care (neither doubt I but yt is yours also) to procure by all meanes possible y[e] floorishing estate of your universitie in religion & liberal sciences:—so at this p[r]sent (to my great comfort) I see yt not inferiour herin to any universitie in Europe or any other pat [*sic*] of y[e] world were yt not y[t] many famous nurseries (as *Padua Montpellier* that of *Vienna* &c.) others had prevented or rather provoked us by their good example, in purchasing of publique gardens and seeking out men of good experience to dresse and keepe the same Whereby that noble science of physicke is made absolute as having recovered ye facultie of *Simpling* a principall and materiall part thereof, wherefore not doubting of your readines in imitating or æmulating the best in so laudable actions I thought yt good to moove you herin & to commend this bearer *Ihon Gerard* a servant of mine vnto you: who by reason of his travaile into farre countries, his great practise & long experience is throughly acquainted with the generall & speciall differences, names, properties & privie markes of thousands of plants & trees. So y[t] if you intend a worke of such emolument to y[r]selves and all young students I shall be glad to have nominated and furnished you with so expert an *Herbarist*: & your selves I trust will think well of the motion and the man Thus desiring god to prosper all your godlie studies and painfull indevors I bidde you hartily farewell."

[Endorsed] "John Gerrard, [an erasure]
A bill [?] of his owne drawing for y[e] L. Ther. [Burleigh] to signe, to y[e] university of Cambridge, for planting of gardens."

(Lansd. MSS. Vol. 107, No. 92, fol. 155.)

---

[15] MSS. Barbers' Hall. [p] Ibid. [16] Ger. pref.
[17] Maitland, i. 978, Noorthouch, 642. Stow, 313., Ibid. ed. Strype, iii. 252.

NOTE.—Timbs has stated that Gerard had a physic garden in Old Street, but I do not know on what authority, nor can I corroborate the statement.—*Vide* Something for everybody, 242 (1861.)

In this year, 1596, our Author made his first appearance in print; being urged by many friends he issued a list of the plants he had cultivated in his own garden, for some years;[18] this catalogue will be found described in the introduction. So far as I can learn, this little work of twenty-four pages is the first professedly complete catalogue of any one garden, either public or private, ever published. There are two previous works indeed of somewhat similar purpose, but as will be seen, they really occupy different ground. In the one case Conrad Gesner drew up a codified list of choice plants, cultivated in the gardens of about twenty of his friends,[19] and short lists follow, of rarities in certain gardens; in the other, Johann Franke, published his Hortus Lusatiæ, an extremely scarce work, in 48 pages, which contains a catalogue of all plants growing near Launitz in Bohemia, both wild or cultivated, the latter being distinguished by the addition of the letter H.

The year following, Gerard was attacked by a "most greeuous ague and of long continuance";[20] subsequently, he was appointed Junior Warden of the Barber-Surgeons,[21] and in December, the work by which his name has been preserved, appeared at his own risk(?),[22] under the title of "The Herball, or generall historie of Plants," etc. The history of this work is curious, and well deserving of attention.

John Norton, the Queen's printer,[23] had commissioned a Dr. Priest, a member of the College of Physicians,[24] to translate Dodoen's Pemptades (1583) from the Latin into English,[25] but the translator dying before the completion of his task,[26] the unfinished work came by some means, into the hands of Gerard.[27] To mask the fact of his Herball being little else than a mere translation, he altered the arrangement from that of Dodoens into that of L'Obel, and flippantly remarking that he had heard of Dr. Priest's labours, but the man being dead, his work had perished with him,[28] he had the effrontery to declare that his own researches had produced the work, to which that statement was prefixed. The wood blocks used by Tabernæmontanus in his Eicones (1590), (not the Neuw Kreuterbuch, 1588), with some others, were procured from Frankfort by Norton,[29] but Gerard soon showed his slender knowledge,[30] by misapplying many of the figures, and caused so much confusion in the early chapters of the Herball, that the attention of the printer was directed to it by James Garret, a London Apothecary, and the correspondent of Charles de l'Escluse. L'Obel was thereupon invited to correct the work, and by his own account he actually corrected it in a thousand places, but further emendation was stopped by the author, who contended that the Herball was already sufficiently accurate, and that his censor had forgotten the English language.[31] I am disposed to credit this assertion, after careful comparison of the names used by Gerard, in both editions of his Catalogus, with those in his Herball, and although L'Obel addressed Gerard in very complimentary terms,[32] yet afterwards he used needlessly bitter language in speaking of his old acquaintance,[33] charging him with pilfering from the Adversaria without acknowledgement,[34] and giving inappropriate names to plants.[35]

The Herball contains upwards of eighteen hundred woodcuts, of which not more than sixteen appear to be original,[36] although Sprengel gives a list of twenty-five, either original or peculiar, some being no improvement upon the old figures;[37] yet Gerard ventured to excuse certain irregularities in his third book, owing to his being "hindered by the slacknesse of the cutters or

---

18 Ger. Cat. ed. I. dedic. 19 V. Cordi, Annot. in Dios. foll. 236—288. 20 Ger. 1006.
21 List at Barbers' Hall. 22 Ger. pref. 23 Lob. Ill. 3. 24 Ger. pref. 25 Lob. Rond. 59.
26 Johns., in Ger. em., pref. 27 Bredwell, lit. in Ger. 28 Ger. pref. 29 Johns., in Ger. em., pref. 30 Lob. Ill. 2.
31 Lob. Ill. 3. 32 Ger. pref., & 55. 33 *Cf.* Lob. Ill. 34. 63. 95, 111. 34 Ib. 95. 35 Ib. 111.
36 Haller, Bibl. Bot. i. 389. 37 Spreng. Hist. i, 466.

grauers of the figures;"[38] the index is very faulty and incomplete. A copy of this work, in the Botanical department of the British Museum, formerly belonging to James Petiver, contains references made by him to Tabernæmontanus, Eicones Stirpium (1590); one hundred and thirty-one, were either from Clusius, or the few original figures above mentioned. This original edition of the work is comparatively seldom quoted, since the emended issue under the editorship of Thomas Johnson in 1633, is greatly superior in every respect; indeed it was almost entirely due to the ability of the editor, that the Herball continued for so long, the standard for English students. It is but fair however to mention, that Gerard modestly avows his own slight attainments,[39] and states that it was principally intended for gentlewomen.[40] Neither of the two editions of the Catalogus, nor the Herball were registered at Stationers' Hall.

Gerard, in Jan. 15, 1598, was appointed one of the examiners of candidates for admission to the freedom of the Barber-Surgeons' Company;[41] later on, an order dated August 1, 1599, was issued by the Queen for the delivery of arms from the City Companies, upon which Master Warden Thornie lent "to Mr. Gerrard, one Corslet and one headpeece, a sworde and a dager." [42] Why he borrowed these weapons, unless for personal defence in that troublous period, I can hardly understand, since the Members of his Craft were exempted by Statute,[43] from being called upon to bear arms, or to serve upon any inquest or watch. The same year witnessed the publication of the second edition of his Catalogus.

I cannot trace any particulars of Gerard during the three years following, at the end of which time he re-appears in the following entry, 2 Nov., 1602. "This day it is ordered that the committee for Mr. Gerrard's garden, shall this afternoon meete at the hall to consider of the report for a garden for the said Mr. Gerrard." [44] No subsequent minute appears with reference to the foregoing, but it is not unlikely, that the action of the committee resulted in a lease being granted to Gerard as under, by the Consort of James I.

[45] "ANNA R.

"Anne by the grace of God Queene of England, Scotland ffraunce and Ireland To ALL, & whome these p'nts shall come greeting, KNOW yee that for and in consideracon of the some of ffive shillings of lawfull money of England in the name of a ffyne to vs before hand payd by John Gerrard of London Surgeon and *Herbarist* to the Kings ma^tie (whereof and wherew^th we acknowledge our self satisfied) as also for divers and sundry other causes and consideracons but especially of his singular and approved art skill and industrie in planting nursing and preserving plants hearbes flowers and fruits of all kindes We are pleased to graunt . . . . vnto the said John Gerrard one garden plot or piece of ground belonginge and adioining on the east part to o^r mansion house called Somersett howse also Strond howse abutting on the west on the wall of the said house on the east vppon the lane comonly called Strond Lane on the south vppon the banke or Wall of the River of Thames and on the north vppon the backside of the Ten^ts standing in the high Streete called the Strond conteyning by estimacon two acres or thereabouts w^h free access ingress . . . . To HAVE AND TO HOLD to the said John Gerrard his executors administrators & assignes from the feast of Saint Michael next ensuing the date hereof the said garden plott or peece of ground and every parte and pcell thereof wth. all and evry the p^rmises and their appurtenences for and during the terme of

[38] Ger. 1077. [39] Ger. pref. [40] Ibid. 707. [41] MSS. Barbers' Hall. [42] Ibid.
[43] 32 Hen. VIII. c. 42. [44] MSS. Barbers' Hall, [45] MSS. Record Office James I. (domestic.) Vol. IX. fol. 113.

oʳ naturall life and for and during the terme of one and Twentie yeares to be accompted and to begin from and ymediately after oʳ decease & fully to be complete and ended YEELDING and paying to vs oʳ executors or assignes during all the terme and termes aforesaide the yearely rent of fower pence of currant money of England to be payd quarterly at the fower usuall feastes . . . . YEELDING also and annswearing yearlie to and for our owne vse onely at the due and proper seasons of the yeare a convenient proportion and quantitie of herbes flowers or fruite renewing or growing wʰin the said Garden plott or piece of grounde by the arte and industrie of the said John Gerrard if they be lawfully required and demanded Given under oʳ seale at Whitehall the ffourteenth daie of *August* in the yeare of the Kings Maᵗⁱᵉ of England ffraunce and Ireland the second and of Scotland the eight and thirtie."

This grant was "endorsed, 30 August, 1604." The draft, which differs from the deed itself in some particulars, was drawn up at Theobalds. Gerard did not long enjoy the use of this garden, for by another endorsement we learn that he parted with all his interest in the lease, 26 Nov. 1605, to Robert Earl of Salisbury, second son of Lord Burleigh, then Lord Treasurer of England and Secretary of State jointly with Sir Francis Walsinghame, until upon the death of the later the whole of the duties of that office devolved upon the former. It is possible that Gerard occupied a similar position in the household of the son as in that of the father.

There certainly must have been some strong reason to prevent his sustaining all the honours of his position, for I find an entry, 26 Sept., 1605,[46] thus, "This day Mr. Gerrard was discharged of the office of second Warden and vpper governor of this Company vppon his suite and entreatie for certayne consideracons, And is fined for the said places x.*l.* ye which he is pʳsentlie to pay pʳsent Mr. or governor And is hereafter to take his place as though he had serued the place of vpper governor Anything to the contrarie notwithstandinge." Gerard consented to make the payment, but subsequently applied to have it remitted; in December he paid the £10,[47] but prayed that it might be treated as a deposit, until the next Court, in the hope that the fine would not be enforced; but the Court held 15 May, 1606, finally decided that the fine could not be foregone.[48] On 21 Oct. 1606, the subject of our memoir was fined by the court, amount not stated, for abusing John Peck, a fellow examiner, and ordered to be friendly and all controversy between them to cease.[49]

In August, 1608, he was elected Master of the Barber-Surgeons' Company, but the books of the Company are missing for that period; consequently I am unable to supply any further details of his life. He died in February, 1611-2, and was buried in St. Andrew's Church, Holborn, on the 18th of that month;[50] but there is nothing to indicate the actual spot; one of his friends Thomas Thornie has an elaborate monument in the Church.

Gerard, no doubt, had as good practical knowledge of plants as any of his countrymen then living, and owing to his patronage by the most powerful statesman of the Elizabethan Court, he had good opportunities of enriching the gardens under his care with new plants, and he certainly cannot be reproached with having neglected those opportunities. His accuracy however was not unimpeachable, he having recorded as natives of this country, many plants he could not have found under the circumstances stated.[51] Johnson and Parkinson, who came into notice, a generation later, were decidedly superior to Gerard, from nearly every point of view. Still Gerard drew

---

[46] MSS. Barbers' Hall. [47] Ibid. [48] Ibid. [49] List of Masters and Wardens, Barbers' Hall.
[50] Regr. St. Andrew's, Holborn. [51] Vide Ger. passim.

attention to indigenous botany, and an impetus was given to the study, which no previous writer had succeeded in accomplishing; a comparison of Lyte's Herball with that under notice, will readily exemplify this. Of his family matters we know next to nothing; he was married,[52] and his wife assisted him professionally, but no hint is given of any other member of his family. The baptismal registers of St. Andrew's, Holborn, which commence in 1558, might throw some light upon this point, but the task would be a long and tedious one, in the total absence of a clue to guide the searcher to any particular period.

The list of names of his acquaintances, more than fifty, scattered through the Herball, is too long to give here. He received plants from all the then accessible parts of the globe, and from men of almost every rank in life. Robin of Paris, previously mentioned, Camerarius of Nuremburg, Lord Zouch, Nicholas Lete, and John Franqueville, the last two London merchants, Thomas Edwards, and James Garret, apothecaries, were amongst the contributors of exotic plants, whilst for indigenous, the names most frequently appearing are Thomas Hesketh, a Lancashire gentlemen, and Stephen Bredwell, a physician. Gerard dispatched one of his assistants, as a ship's surgeon to the Mediterranean, in the Hercules,[53] that he might bring home some new plants. He, himself, had travelled over a large part of England, but Salmon's statement as to his living in Lincolnshire[54] refers to Johnson. There is no will of Gerard's at Somerset House, but it is not probable that he acquired wealth; the printer of the Herball in this respect, was more successful than the compiler.

A half length portrait of Gerard, engraved by William Rogers, faces p. 1. of the Herball; he holds a branch of the Potato plant. Beneath are his own arms, those of the City of London, and of the Company of Barber-Surgeons.

A reduced copy of this portrait appears on the title page of Johnson's edition, and Sir J. E. Smith possessed a copper plate[55] of another engraved by Hall, much worn, but I have not succeeded in tracing it, nor have I seen any impression from it.

[52] Ger. 695.

[53] Ger. 1304.

[54] Salm. Herb. i. 64.

[55] Rees' Cyc. art. Gerard.

Catalogus arbo-

rum, fruticum ac plantarum tam
indigenarum, quam exoticarum,
in horto Iohannis Gerardi *ciuis*
& Chirurgi Londinenſis
naſcentium

[The Royal Arms,
in a garter.]

LONDINI
Ex officina Roberti Robinſon
1596.

[Reprinted, 1876.]

[The Arms and Supporters of
Lord Burleigh, with his motto
COR VNVM, VIA VNA.]

Honoratiſs : atque pru-

dentiss : viro Domino G. Cecilio Baroni

Burleienſi : Nobili Equeſt. ordinis So-

dali Reg. Conſilij Senat. grauiſs. Sum-

moque Ang. Theſaurario, &c.

Fœlicitatem optat, I. G.

S*Vaſerunt ſæpius amicorum plurimi, rei Herbariæ ſtudioſi (Honoratiſſime D.) vt Stirpiū ſiue Herbarum, quas meo non vulgari ſtudio et induſtria, ex remotſs. partibus quæſitas, non ſegni cura et labore, in Hortulo meo ſuburbāno per aliquot annos coluiſſem; Catalogum aliquē in publicum darē. Horum ego precibus tandem victus, (quibus negare nefas eſſet) Catalogum hunc manu propria, non sine molestia, describebam, quem cum illis communicaſſem, ita mecum egerunt; vt nisi in lucem emitterem, vix illis ſatiſfacerem. Gratior ſcio ſtudioſis nostra opera futura eſſet; ſi non catalogum modo earum ſtirpium, quas apud*

apud me iamdiu alui, ſed etiam icones natiuis ſuis coloribus depictas, ſuiſque abditis virtutibus ornatas, nostro idiomate euulgaſſem: Cuiusmodi opus iampridem meditatum, nunc ferè abſolutum, veluti foetum partui vicinum, tempus in lucem proferet. Accipe interea (illuſtriſs. Domine,) a ſeruo tuo, leuiusculum hoc munus, maioris nostri operis præludium; tuoque benigno aſpectu faſciculum hunc dignare, ſicut ſemper hactenus dignatus es; vt inde non minus foueatur, quàm flores Solis radiis reficiuntur. Ita fiet vt non ſolum præteriti laboris et industriæ non me vnquam poenitebit, ſed etiam futuri non pigebit, Et si quid emolumenti inde recipiant studiosi rei herbariæ, Tuæ D. acceptum ferant.

Tibi semper deuinctiss.

Joh: Gerardus.

Perbonis & studiosis stripium indaga-*toribus. Io: Gerardus.*

OMnes hoc iucundissimo studio captos, rogatos velim, vt si quas præter has plantas reperiant; easdem nobis liberaliter communicent & nostros conatus iuuent, sibique persuadeant tanto & reciprocomunere impertiri.

Catalogus Horti Iohannis Gerardi *Londinenſis*

ABies
Abrotanum mas
Abrotanum foemina
Abutilon Auicennæ
Abſinthium latifolium
Abſinthium Santonicum
Abſinthium marinum
Abſinthium folio Spicæ
Abſinthium inſipidum
Acanthus ſatiuus
Acanthus Germanicus
Acatia prior Matthioli
Acatia ſecund Matth:
Acer Maior
Aconitum hiemale
Aconitum Delphinias
Aconitum luteum ponticũ
Aconitum lycoctonon
Aconitum ſeu Napellus
Aconitum folio Platani
Acorus verus
Adianthum nigrum
Æthiopis
Agrioriganum
Album olus
Alcea arborea
Alcea Veneta
Alcea fruticoſa pẽtaphyllea
Alchimilla
Allium flore luteo
Allium vrſinum latifolium
Allium vrſinu anguſtifoliũ
Allium proliferum
Alliaria
Allyſon Dios:
Alnus nigra
Aloe
Alopecuros
Alleluia
Alſine repens
Alſine foliis Triſſaginis
Althea arborea
Althea floribus luteis
Amara dulcis
Amara dulcis flore albo
Amaracus
Amaranthus purpurens
Amaranthus maior
Amaranthus tricolor
Ammi vulgatius
Ammi Creticum

Amigd.

Amigdalus arbor
Amomum Plinii
Ampeloptaſſon
Anblatum
Anagallis flore cœrnleo
Anagallis flore phoeniceo
Anagallis flore luteo
Anagyris
Anchuſa
Anchuſa Neapolitana
Androſaces Math:
Androſemum
Anemone maxima polyanthos (plex
Anemone coccinea multi-
Anemone coccinea ſimplici flore
Anemone Gerani-folia
Anemone tuberoſa tadice
Anemone tenui-folia ſimplici flore
Anemone tenui-folia flore coeruleo
Anemone tenui-folia flore albo
Anemone flore luteo
Anemone ſylueſt
Angelica ſatiua
Angelica ſylueſtris
Aniſsum
Anteuphorbium
Anthemis duplici flore
Anthora

Antirhinnm album
Antirhinum purpureum
Anthillis leguminoſa
Apyos Fuchſii
Apocynum rectum
Apocynum repens
Aquilegiæ variæ
Arabis
Aracus
Arbor Iudæ
Arbor Vitæ
Armeria ſylueſtris
Armeria alba
Armeria rubra mul.iplex
Armeria polyanthos
Armeria alba guttata
Armeria ſuaue flore rubēte
Armeria prolifera
Argemone Taberne motana
Ariſarum latifolium
Ariſarum anguſtifolium
Ariſtolochia longa vera
Ariſtolochia totunda vera
Ariſtolochia clematitis
Artemiſia, mater herbarum
Artemiſia leptaphyllon
Artemiſia marina
Aſarina
Aſarum Baccaris
Aſclepias
Aſclepias flore nigro
Aſcyron Creticum
Aſperula coerulea

Afphodelus albus
Afphodelus albus ramofus
Afphodelus Bulbofus
Afphodelus fiftulofus
Afphodelus Luteus
Afphodelus lancaftriænfis
Afparagus
Afplenium
After atticus flore luteo
After atticus flore cæruleo
After Inguinalis
Aftragalus
Aftragaloides
Aftrantia
Aftrantia nigra
Afureus conuoluulus
Atriplex horten: Alba
Atriplex horten: rubra
Atriplex olida
Atractylis
Atractylis hyrfutior
Auricula vrfi flore purpureo
Auricula vrfi flore luteo
Auricula vrfi flore vario
Auena nuda

B

BAmia
Balfamina cucumerina
Balfamina foemina
Balfamita mas
Balfamita foemina
Barbarea
Barba capri Fuchfii
Bellides variæ
Bellis prolyfera
Behen album
Behen rubrum
Berberis maximo fructu
Berberis fine acinis
Beta rubra
Beta nigra
Beta alba
Betonica flore albo
Biftorta maior
Biftorta maior altera
Biftorta minor
Blattaria flore luteo
Blattaria flore purpureo
Blattaria flore albo
Blattaria flore rubro
Blitum album
Blitum rubrum
Blitum fupinum
Bolbocaftanon
Bonus Henricus
Borago femper virens
Botrys
Braffica florida
Braffica fimbriata
Braffica marina monofpermon
Braffica tricolor
Braffica caulirapa
Braffica patula
Braffica arboreffens

Braſſica exotica
Bugula flore albo
Bulbus eryophorus
Buphthalmus verus Do.

C

Cacris vera
Cakyle ſerapionis
Calamenta motana
Calamenta preſtantior
Calendulæ variæ
Camædrys
Camædrys laciniatis foliis
Campanula lactefcens
Campanula perſici-folia
Campa: Perſicifolia alba
Canna Indica
Capnos fabacea
Capnos alba
Capparis vera
Capparis leguminoſa
Capſicum Actuarii
Capſicum Indicum
Cardiaca
Cardiaca ſpinoſa Camerarii
Carduus ſtellatus
Carduus a Caulis
Carduus tomentoſus
Carduus globoſus
Carduus Gerardi
Caryophyllorum hortenſiuum variæ in colore diffe-
Caryophyllus flore luteo
Caryophyllata Alpina
Caryophyllata rotundifolia
Carum
Caſtanea
Catanance
Caucalis Hyſpanica
Caucalis Cræatenſis
Caucaſon
Cauda muris
Centaureum flore albo
Centaureum luteum Lobelii
Centau. magnum
Ceraſa Anglica ſerotina
Ceraſa Belgica
Ceraſa alba Hiſpanica
Ceraſa racemoſa
Ceraſa agriotta
Ceraſa ſerotina altera
Ceraſa Gaſconica
Ceraſa cordata maiora
Ceraſa cordata minora
Ceraſa nigra maiora
Ceraſa nigra minora
Ceraſa duplici flore
Ceraſa duplici flore altera
Ceraſa coerulea
Ceratia ſiliqua
Cerinthe Plinii
Cerinthe maius
Ceruicaria maior
Ceruicaria minor
Ceruicaria Gerardi
Chamæficus
Chamæ ceraſus Alpigena

che-

Chamælea tricoccos
Chamælea
Chamælinum puſillum
Chamælea alpina glauca, ar-
genteaue
Chamæmorus
Chamæpytis
Chamænerium
Chamæiris flore rubello
Chamæiris lutea
Chamæiris niuea
Chamæiris purpurea
Chamæiris variegata
Chamæiris auguſti folia
Chamæiris violacea
Chamæiris latifolia
Chamæiris variegata Cluſii
Chriſtophoriana
Chryſanthemum proliferum
Chryſanthemum Peruuianum
Chryſanthemum aruorum
Circæa
Cirſium
Ciſtus mas
Ciſtus foemina
Ciſtus humilis
Cytiſus Maranthæ (bo
Clematis peregrina flore al-
Clematis pere : flore rubro
Clematis Boetica
Clematis Pannonica
Clematis Daphonides
Clynopodium

Climenum Italorum
Cnicus ſatiuus
Cochlearia Britanica
Cochlearia Batauorum
Colchicum Anglicum albū
Colchicum Pannonicum
Colchicum luteum
Colchicum ephemerum
Colus Iouis
Colutea
Colutea ſcorpioides
Colutea minima
Condrylla rara flore purpurate
Condrilla flore coeruleo
Condrylla flore luteo
Coniza maior
Conizæ variæ
Conſolida media vulnerarioru
Conſolida segetum
Conſolidæ regales variæ
Conuoluuli varii
Coriandrum
Cornus mas
Cornus fructu albo
Cornus foemina
Coronopus
Coronopus Ruellii
Corona Imperialis
Corona Terræ
Cortuſa Mathioli
Cotyledon
Chamæmalus

Craſſula maior
Crateogonon
Corylus Tripotitanus maximus
Crocus Anglicus
Crocus montanus
Crocus vernus flore albo
Crocus vernus flore luteo
Crocus vernus flore violacea
Crocus vernus flore vario
Crutiata herba
Crutiata gentiana
Cucumer Aſininus
Cucurbitæ variæ
Cuminum ſatiuum
Cupreſſus
Cyanus maior
Cyanus varia genera
Cyclamen ſolio Hederæ
Cyclamen orbiculato folio
Cynara
Cynocrambe
Cynogloſſum
Cynogloſſum puſillum
Cynogloſſum cræticum

D

DActylo prlunum
Daucus cræticus
Daucus ſelinoides
Dens caninus
Dentaria maior
Dentaria alabeſtritis
Dentillaria Rondeletii
Digitalis alba
Digitalis flore luteo
Digitalis purpurea
Dictamnum cræticum
Doronicum Romanum
Draba vera
Draba altera
Draco herba
Dryopteris

E

EBulus
Elatine
Elatine foemina
Elaphoboſcum verum
Elleborine
Epymedium
Eringium marinum
Eringium mediteraneum
Eringium planum
Eruca peregrina
Eruca naſturtio cognata
Eſula maior Hiſpanica
Eſula minor
Eſula exigua
Eſula rotunda.
Eupatorium Avicennæ
Euonymos Theophraſti

F

Fabia.

Faba grecorum
Fabæ variæ
Ferula galbanifera
Ferula ſagapenifera
Ferula nigra
Ferulago
Ficus de Algara
Ficus indica
Filix florida
Filix mas
Filix foemina
Flammula
Filipendula
Flos adonis
Flos africanus maior
Flos africanus minor
Flos aſcricanus ſimplex
Flos ſolis
Foenum Burgundiacum
Ferrum equinum
Fragaria ſterilis
Fragaria rubra
Fragaria alba
Fragaria ſubuiridis
Frangula
Fraxinus bubula
Fraxinella
Frittillaria
Fumaria alba
Fumaria lutea
Fumaria latifolia

G

GAlega
Caleopſis pannonica
Gallium allbum
Gallium luteum
Geniſta Hiſpanica
Geniſtella
Gelſeminum album
Gentiana maior
Gentiana Gerardi Anglica
Gentianella
Geranium batrachioides
Geranium bulboſum
Geranium Cræticum
Geranium fuſcum
Geranium gruinum
Gera malacoides
Geranium repens
Geranium Robertianum
Geranium flore albo
Geranium flore cæruleo
Geranium columbinum
Geranium non ſcriptum
Geranium moſchatum
Gingidium
Gladiolus Narbonenſis
Gladiolus Italicus
Gladiolus flore pallido
Glaſtum
Glaux dioſco ridis
Glaux exigua
Glaux vulgaris
Glicyrrhiza ſiliquoſa
Glicirrhiza echinata

Gnaphalium montanum
Gnaphalium marinum
Gnaphalium Anglicum
Gnaphalium Americum
Gramen Parnaſi
Gramen ſtriatum album
Gratiola
Gratiola Gerardi Anglica
Guaiacum Patauinum

H

HArmala
Halicacabum
Halymus
Hedyſarum
Hedyſarum clypeatum
Hedypnois
Helleboraſtrum vtrunque
Helleborine radice repente
Helleborus niger verus
Helleborus niger alter
Helleborus albus
Helleborus albus atrorubens
Helenium
Helxine
Helxine ciſſampelos
Hemerocallis Valentina
Hemionitis ſterilis
Hepatica nobilis flore albo
Hepa : nobilis flore rubro
Hepa : no : flore coeruleo
Herba Doria
Herba Paris
Herba Turca
Herba Gerardi
Herba venti Rondeletii
Hermaphroditica orchis
Hermionitis ſterilis
Hermodactylus Italorum
Hieratium grandius
Horminum verum
Horminum ſylueſtre
Horminum hortenſe
Hyacinthus Anglicus cæruleus
Hya : Anglicus albus
Hya : Ang : ſuaue rubens
Hyacinthus autumnalis
Hyacin botroides
Hya : botroides albus
Hya : botro : amoenus
Hya : orientalis ceruleus
Hya : orient : albus
Hya orient Græcus
Hya orient brumalis
Hya ſtellatus fuchsii
Hya : ſtell : Byzantinus
Hya : ſtellat : Germanicus
Hya. comoſus maior
Hya comoſ. minor
Hya : como Byzantinus
Hya : como. albus
Hyoſchiamus albus
Hyoſchiamus niger
Hyoſch : lutens
Hypecoon Clusii
Hyppogloſſum Bonifacia

Hys-

Hyſſopus flore albo
Hyſſopus Gerardi
Hyſſopus lati-folius
Hyſſopus Criſpus
Hyſſopus Cræticus
Hyſſopus niueus Anglicus
Hypolapathum rotundi-folium

I

IAcea maior flore purpureo
Iacea maior flore luteo
Iacea maior flore flauo altera
Illicebra
Iris biflora Luſitanica
Iris Florentina
Iris Dalmatica maior pallida & coerulea
Iris Dalmatica minor
Iris ſyluestris Byzantina peramæna
Iris maritima Narbonenſiū
Iris Narbonenſis minor
Iris variegata Clusii
Iris violacea parua
Iris Calcedonica variegata
Iris obsoleto flore
Iris noſtras paluſtris
Iris Suſiana
Iris purpureo flore
Iris bulboſa flore coeruleo
Iris bulboſa flore luteo
Iris bulboſa flore vario
Iris bulboſa varia altera
Iucca, Indiæ occidentalis planta quæ alia à Yuca Indorum, exqua panisfit, videtur; nam quamuis foliis ſit petpetuò virentibus, iiſque minime laciniatis; ſed ambitu, Draconis arboris inſtar ex atro rubentibus, mucronatis & peracutis; quodam mado Sedum aquaticum Belgarum preſeferentibus; Radix ſubeſt Aſphodeli; Poeniæ foeminæ par & concolor

K

KAli magnum
Kali minus
Keyri multiplex varietas
Knawel ſiue ſaxifraga altera Anglica Lobelii

L

LAcrima Iobi
Lactucæ variæ
Lactuca ſylueſtris ſoporifera
Lagopus
Lagopus maximus
Lamium album

Lamium luteum
Lamium pannonicum
Lampſana
Lanaria herba
Lathyrus anguſtifolia
Lathyrus lati-folia
Laurus Tynus
Lens
Lepidium
Leucoium bulboſum precox maius
Leucoium bul: precox minus
Leucoium bulb hexaphyllon
Leucoium triphyllon
Leucoium marinum
Leucoium luteum multiplex
Leuiſticum
Licium Italicum
Lilium non bulboſum luteum
Lilium non bulboſum phænicoeum
Lilium Alexandrinum
Lilium bizantinum
Lilium montanum
Lilium rubrum
Lilium album
Lilium album bizantinum
Lilium Perſicum
Lilium cruentum
Lilium cruentum bulbigerum
Lilium conuallium flore rubello

Limonium magnum
Limonium paruum
Linaria aurea
Linaria valentina
Linum ſelueſtre
Linum marinum
Lotus tetragonolobus
Lotus vrbanus
Lotus arbor
Lunaria, bolbanac
Lunaria magorum
Lunaria raphanitis
Lunaria minor
Lupinus ſativus
Lupinus flore luteo
Lupinus flore coeruleo
Lycopſis
Lichnis agreſtis multiflora alba (rubra
Lychnis agreſtis multiflora
Lichnis marina Anglica
Lichnis coronaria alba
Lychnis coronaria rubra multiplex
Lychnis calcidonica
Lylac mathioli
Lyſimachia lutea
Lyſimachia flore coeruleo
Lyſimachia ſiliquoſa
Lyſimachia ſpicata
Lyſimachia galericulata
Lythoſpermum maius
Lythoſpermum minus

M

MAla insana
Mala insana altera
Mali persici decem varietates
Malus Armeniaca
Malua Geranifolia
Malua crispa
Malua arborescens coccinei coloris
Malua arborea polyanthos rubro flore
Maluæ arboreæ variæ
Malum punicum
Marum
Marrubium album
Marubium Cræticum
Martagon Chymistarum
Martagon imperiale
Matricaria grato adore
Matricaria duplici flore
Medica
Medica spinosa
Medica Arabica
Medica marina
Melampyrum
Melanthium Damascenum
Melanthium flore luteo
Melanthium flore albo
Melant: pleno flore alterum
Melilotus coronata
Melilotus Germanica vtraq,
Melilotus Italica
Melilotus Arabica
Mentæ variæ
Melissa
Melissa Turcica
Melissa Moluca
Melones saccharati varii
Melocoton
Meon
Mercurialis mas
Mercurialis foemina
Mespylus sativus
Morsus gallinæ
Morsus gallinæ hedæraceus
Mezereon
Millefolium legittimum
Millefolium rubrum
Millifolium album
Millium
Millium Indicum
Mirabilia Peruuiana
Morus alba
Morus rubra
Moluca spinosa
Moli Dioscorideum
Moli Homericum
Moli Indicum
Moli serpentinum
Moli foliis Narcissi
Moli montanum latifolium
Mollugo
Momordica

Monophyllon
Morion Theophrafti
Mufcari flore luteo
Mufcari cineritium
Mufcipula
Mufcipula vera
Mitulo Prunum, fiue Prunum Mituli effigie
Myrhida Plinii
Myrrhis
Myrtus Brabantica
Myrtacantha

N

NArciffus medio luteus
Narciffus medio purpurens
Narciffus medio pur: precox
Narciffus medio purpureus precotior
Narciffus minor ferotinus
Narciffus Pifanus
Narciffus albus Bizantinus multiplex
Narciffus albus Germanicus multiplex
Narciffus luteus multiplex
Naciffus Perficus
Narciffus Iunci-folius
Narciffus totus luteus
Nafturtium Indicum
Nidus Auis
Nummularia
Nux Iuglans
Nux veficaria

O

OCymum maximum
Ocymum minimum
Ocymoides
Oenanthe aquatica
Oenanthe cicutæ faciæ
Oleander
Oleafter
Ononis flore albo
Ononis non fpinofa
Ophyogloffum
Ophyofcorodon
Orchis andrachnitis
Orchis melitias
Orchis ornithophora
Orchis apifera
Orchis fpiralis
Orchis radice repente
Ornithogalum
Ornithogalum Pannonicnm
Ornithogalum luteum
Ornithopodium
Origanum Cræticum
Orobus
Ofmunda
Othonna polyanthos
Oxalis rotundi-folia

P

PAliurus
Panax Chyronium
Panax Afclepium
Panax Heracleum
Panax Gerardi Mentastri-folia
Panicum album
Panicum rubrum Americanum
Papauer fimplex purpureo flore
Papauer fimplex flore albo
Papauer polyanthos rubro flore
Papauer polyanthos albo flore
Papauer corniculatum flore luteo
Papauer corniculatum phoeniceo flore
Papus orbiculatus
Papus Hyfpanorum
Paronychia alfine folia
Paronychia rutaceo folio
Parthenium Alpinum
Pæonia mas
Pæonia foemina
Pæonia polyanthos
Pæonia promifcua
Pæonia albicans
Pecten Veneris
Pentaphyllum maximum
Pentaphyllum album
Pentaphyllum rubrum
Peplis
Peplios
Perfoliata
Perfoliata filiquofa
Periclymenum
Periclymenum perfoliatum
Periclymenum arborefcens
Perchepier Anglorum
Petafites
Petrofelinum Macedonicum verum
Petrofelinum crifpum & conplicatum
Peucedanum
Phalangium ramofum
Phalangium non ramofum
Phalaris
Phafeoli varii
Phyllitis
Phyllitis multifido folio
Picea
Pimpinella
Pinguicula
Pinus
Pinafter
Pifum cordatum
Pifum vmbelliferum
Pifum excorticatum
Pifū minus ex luteo virescēs

Pisum perenne
Plantago rosea
Plantago rosea Gerardi
Plantago marina
Platanus verus
Polemonium
Polium montanum
Polygala flore albo
Polygala flore coeruleo
Polygala rubens
Polygonatum
Polygonatum Pannonicum
Polygonatum minus
Polygoni varia genera
Poma amoris rubro fructu
Poma amoris flaua
Poma Ægyptia
Pomum spinosum
Poterion
Primula veris flore rubro
Primula veris viridi flore
Primula veris viridi multiplici flore
Primula veris flore geminato
Primula veris maxima Anglica
Primulæ syluarum variæ
Prunella flore albo
Pruni arboris species 30
Pseudo-dictamnum
Pseudo-costus
Psyllium
Psyllium semper virens
Pseudo-narcissus
Pseudo-narcissus Hyspanicus
Pseudo-bunium
Ptarmica
Ptarmica duplici flore
Pulegium erectum
Pulegium regale supinum
Pulmonaria vera
Pulsatilla
Pyrethrum officinarum
Pyrola

Q

Quadrifolium phoeum
Quinqueneruia rosea

R

Radix caua flore purpureo
Radix caua flore albo
Radix caua viridi flore
Ranunculus Alpinus
Ranunculus magnus Anglicus Polyanthos
Ranunculus bulbosus
Ranunculus Illiricus
Ranunculus niueus Polyanthos
Ranunculus Gramineus
Ranunculus globosus

Ranuncu-

Ranunculus Tripolitanus
Ranunculus echinatus
Raphanus
Raphanus niger
Raphanus pyriformis
Raphanus ruſticanus
Rhamnus
Rhaharbarum monachorum
Rheſeda Plinii
Rheſeda maior
Rhodia radix
Rhus ſiue ſumach
Ribes nigra
Ribes alba
Ribes rubra
Ricinus
Roſa Anglica alba ſimplici flore
Roſa Anglica alba multiplex
Roſa rubra
Roſa rubra flore maximo
Rosa Damaſcena flore multiplici
Roſa prouincialis
Roſa moſchata ſimplici flore
Roſa moſchata multiplex
Roſa moſchata Hiſpanica
Roſa holoſericea
Roſa lutea
Roſa pomifera
Roſa canina
Roſa canina multiplex odorata
Roſa cinnamomea (tiplici
Roſa cinnamomea flore mul-
Roſmarinum
Roſmarinum cachriferum
Rubus Idæus
Rubus ſaxatilis
Rubia ſatiua
Rubia ſylueſtris
Rubia aquatica
Ruta ſatiua
Ruta ſylueſtris
Ruta aquatica
Ruta muraria

S

Sabdrariffa
Sabina vulgaris
Sabina baccifera
Salix roſea Gerardi
Salicornia
Saginæ Spergula
Saluia flore albo pinnata
Saluia baccifera
Saluia maculata
Saluia criſtatis oris
ſaluia auriculata
Saluia Italica flore candido aromatico, partim folio vulgaris
Saluiæ, minore parrim pinnato
Sambucus montana racemoſa
Sambucus roſea
Sambucus aquatica
Sambucus laciniatis foliis

Sanguiſorba
Sanicula vulgaris
Sanicula guttata
Saponaria
Satureia vera
Saxifraga alba
Saxifraga aurea
Saxifraga Anglica
Securidaca
Scabioſa peregrina
Scabioſa maior Hiſpanica
Scabioſa flore rubro
Scabioſa marina
Scamonium Monſpelienſium
Scamonium Syriacum verum
Schoenopraſſon
Scordium
Scordothlaſpi
Scorodo praſſon
Scorzonera
Scrophularia
Scrophularia Indica
Scorpioides Dodonei
Scorpioides Mathioli
Scropioides bupleurifolio
Scorpioides
Sedum maius
Seriphium
Serpentaria maior
Serpillum
Serpillum Pannonicum
Serratula
Serratula flore albo

Seſamoides magnum
Seſamoides paruum
Seſeli Æthiopicum frutex
Seſeli Cræticum
Seſeli pratenſe
Seſeli peloponenſe
Siſarum
Siſon
Sida marina
Siciliana
Smirnium Cræticum
Soldanella
Solidago ſaracenica
Solanum hortenſe
Solanum ſomniferum
Solanum læthale
Solani ſomniteri ſimilis fruticoſa ignota planta, ſemine Conſtantinopolitano oriunda & delata a nobililiſs. viro domine baroni Eduardo Zouche, foliis tamen rotundioribus, & aliquantulum cauis.
Sorbus torminalis
Sorbus Alpina
Sorbus ſilueſtris
Sophia chirurgorum
ſorghum
ſpeculum Veneris
Spondylium
Staphis agria
Stachis odorata
Stachis Monſpelienſis

Stoebe

Stoebe Salamantica
Stoebe argentea
Stoechas Arabica
Stoechas nudis cauliculis
Superba Auſtriaca
Superba pratenſis
Superba duplici flore
Staphylinus Crætica
Stramonium peregrinum Lobelii
Symphitum magnum
Symphitum tuberoſum
Symphitum petreum
Syringa Italica

T

TAmariſcus Germanicus aut Nerbonenſis
Tamariſcus Italicus
Tanacetum criſpum
Tanacetum inodorum
Tapſus barbatus
Taxus
Telephium ſemper virens
Telephium magnum Hyspanicum
Teſticulus odoratus
Teucrium
Thalictrum magnum
Thalictrum paruum
Thlaſpi Candiæ
Thlaſpi minus
Thlaſpi clypeatum
Thlaſpi vmbellatum
Thymum legittimum
Thymus durius
Thymum durius alterum suauiſſimum
Thapſia
Tithymalus paralius
Tithymalus charatias
Tithymalus mirtifolius
Tithymalus cypariſſias
Tithymalus dedrioides
Tithymalus tuberoſus
Tilia
Tormentilla
Tordilion
Trachelium magnum
Trachelium minus
Tragopogon luteum
Tragopogon purpureum
Tragos
Tragoriganum
Tragium Grmanicum
Tragium Bellonii
Tribulus terreſtris
Trifolium bituminoſum
Trifolium fuſcum
Trifolium Briſtolienſis
Triorchis lutea
Tripolium magnum
Tripolium paruum
Tulipæ infinitæ

V

VAccaria
Vaccinia nigra

Vaccinia alba
Vaccinia rubra
Valeriana maior
Valeriana Græca
Valeriana Indica
Valeriana rubra
Valeriana aquatica
Verbaſcum Matthioli
Verbaſcū matthioli odoratū
Verbaſcum Cræticum
Verbaſcum nigrum
Verbaſcum foemina
Verbaſcum album
Veronica mas
Veronica recta Pannonica
Veronica foemina
Vincaperuinca
Vincaperuinca flore albo
Vincaperuinca flore purpureo
Viola Mariana
Viola calathiana
Viola Theophraſti
Viola Hyſpanica
Viola Matronalis
Violæ Martiæ variæ
Viurna
Virga aurea
Vitex
Vites viniferæ variæ
Vitis alba
Vitis nigra
Vmbilicus Veneris
Vrtica Romana
Vua criſpa baccis rubris
Vua Zibeba

X

XYris
Xanthium
Xylon
Zyziphus

H*Erbas, ſtirpes, frutices, ſubfrutices & arhuſculas hoc Catalogo recenſitas, quamplurimas ac fere omnes me vidiſſe Londini in horto* Iohannis Gerardi *Chirurgi & botanici per-optimi, (non enim omnes eode ſed varijs temporibus anni pullulaſcunt, enaſcuntur aut florent) atteſtor*

Matthias de Lobel

Ipſis Calendis Iunij, M. D. XCVI.

CATALOGVS ARBORVM, FRVTICVM AC PLANTARUM TAM INDIGENARVM, QUAM EXOTICARVM,

in horto *Johannis Gerardi* Ciuis
& chirurgi Londinenſis
naſcentium.

[The printer's device, consisting of two cornucopiæ combined with the symbols of Mercury, in front of a landscape.]

LONDINI.
Ex officina Arnoldi Hatfield,
impenſis Ioannis Norton.
1599.

The verso contains the armorial bearings of Sir Walter Raleigh, with the motto AMORE ET VIRTUTE, and beneath it,

Quid iuuat aut Arcton vidiſſe aſtrumve Canopi,
Ni, qui admirentur ſuſpiciantque, habeas?

CLARISSIMO, ET AMPLISSIMO
VIRO DOMINO WALTERO RALEGH, EQVITI
aurato, rei metallicæ per Cornubiam & Deuoniam
Præfecto, Ducatus Cornubiæ & Exoniæ Sene-
ſchallo, & Regij ſatellitij Capitaneo, Domino
ſuo plurimum obſeruando.

CVm me genius meus, & medicæ artis pars antiquiſſima (cui à teneris ſum initiatus) Chirurgia, in admirationem primum, mox in ſerium rei herbariæ ſtudium impuliſſent; ita me tum honeſta voluptas deliniuit, tum vtilitas immenſa deliniuit, vt nec illius pœniteat, nec huic non plurimum debere me agnoſcam: quare conquiſiui vndique, quicquid Anglia noſtra ſuppeditat, alieno etiam orbe magnis ſumptibus accerſita, aſſiduitate infracta noſtrorum hortorum curam lubenter ſubire condoceferi. Hanc ego ſupellectilem longè antè mihi comparabam, fouebam, cum Botanologicum noſtrum nuper foras datum meditarer. Ex quo opere ad noſtrates non mediocre, vt exiſtimo, redijt commodum, adeoq; ex ijs multos, tanquam claſſico ſigno dato, incendi, meoq; exemplo ad præclarum hoc ſtudium inuitaui: ſic vt nec Honorem tuum, Mæcenas nobiliſſime, Napæas dedignari facilè, colligam. Id quod mihi tanquam palmarium obtigiſſe duco. Quid enim gratius ampluiſq; potuiſſet vnquam mihi accidere, quam in famulitium & clientelam tuam cooptari? Cuius tam domi moderatio animi, & in rebus gerendis dexteritas, quam longinquis partum expeditionibus peregrinationibuſq; nomen, in magnam ſpem Angliam vniuerſam erexit, vt de tuo Honore, accreſcentibus annis, ſumma ſibi polliceatur. Quidni igitur, ego me omni obſequio, tali viro deuoueam & mancupio addicam? addico ſanè, ac lubens, & opellam noſtram quantulamcunque vbicunque humillimè ſubiectam profiteor. Atque adeo, quicquid ego à plurimis iam retrò annis ſtirpium ſiue exoticarum ſiue indigenarum alo, (alo autem vtriuſque generis & plurimas & rariſſimas) ſiquidem præſentius mihi hoc tempore non occurrit ſtudij mei & deuotiſſimi obſequij teſtandi pignus, tui facis iuris, vt Honoris tui arbitrio dominum mutent: vel eo nomine gratæ, vt ſpero, futuræ, quod cum eas cominus in horto tuo quotidie contemplabere, Indicarum nauigationum tuarum, rerumq; orbe remotiſſimo geſtarum dulciſſimam memoriam refricabunt. Cæterum vt Honori tuo rectius de munuſculo conſtet, Catalogum hunc, eccum! concinnaui, præambulonemq; offero, cuius indicio & quod voles petas, & quod petes habeas. Nec enim, ſcio, Patronum dedignabuntur Napææ eum, a quo harum dominus vnice pendet.

T. H. omni obſequio deuinctiſs.
cliens & ſeruus,
JOHANNES GERARDVS.

Rei herbariæ studiosis salutem.

NOn paucos horti nostri celebritas perpulit vt serio apud me sæpius de Indice plantarum, quas is alit, tam nostro cælo natas, quam aliunde petitas plurimas & rarissimas, instarent: quibus cum graue esset tam honesta petentibus operam hanc denegare, longe autem grauiorem & molestiorem futurum prospicerem tam crebro repetitum exscribendi laborem, iccirco tum ne amicorum precibus deessem, & ego simul & semel tædio isto defungerer, prælo horti mei Catalogum subieci. Quod si quid viri φιλοβότανοι, *vel fortunâ bonâ repertum, vel labore improbo comparatum in hoc genere habetis, rogatos volo vt communicetis, mutuam liberalitatem nostram experturi.*

CATALOGVS HORTI

IOHANNIS GERARDI

CIVIS ET CHIRVRGI

Londinensis.

[WITH REFERENCES TO GERARD'S *Herball* (1597) AND MODERN NAMES APPENDED.]

Abies The Firre tree. 1181. **Pinus Abies**, *L.*

Abrotanum mas The male Southernwood. 947, 2. **Artemisia Abrotanum**, *L.*

A. fœmina Female Southernwood. 947, 1, *descr. not the fig.* *Ger. em.* 1105, 1, **A. arborescens**, *L.*

Abutilon Auicennæ Purple [?] mallow. 790. **Sida Abutilon**, *L.*

Absinthium latifolium Broad leafed Wormwood. 940, 2. **Artemisia rupestris**, *L.*

A. Santonicum Holly Wormwood. 941. **A. Santonicum**, *L.*

A. marinum Sea Wormwood. 944. *Ger. em.* 1102, 4. **Santolina Chamæcyparissus**, *L.*

A. folio Spicæ Wormwood with leaves like Lauender. 946, 3. **Artemisia cærulescens**, *L.*

A. insipidum Vnsauorie Wormwood, 943, 4. **A. Absinthium**, *L. var.* = **A. inodora**, *Mill.*

Acanthus satiuus Garden Beares breech. 986. **Acanthus mollis**, *L.*

A. Germanicus Cow Parsnep, or medow Parsnep. 855, *descr.* **Heracleum Sphondylium**, *L.*

Acatia prior Matthioli The Aegyptian Thorne of Matthiolus. 1149, *descr. only.* **Acacia vera**, *Willd.*

A. secunda Matthioli The Thorne of Aegypt. ["Dubium." Lob. MS.] 1149. **Cytisus spinosus**, *Lam.*

Acer maior The great Maple. 1300, *descr. not the fig.* **Acer Pseudoplatanus**, *L.*

Aconitum hiemale Winter Wolfesbane. 819. **Eranthis hyemalis**, *Salisb.*

A Delphini Larks heele Wolfesbane. 822, 1. **Delphinium elatum**, *Ait.*

A. luteum ponticum Yellow Wolfes bane. 821. **Aconitum Lycoctonum**, *L.*

A. lycoctonon Deadly Wolfes bane. 822, 2. **A. variegatum**, *L.*

A. seu Napellus Wolfesbane with the Turnep roote. 823, 3. **A. Napellus**, *L.*

A. folio Platani Wolfes bane with the Plane tree leafe. 823, *descr.* *Ger. em.* 972. 4. **A. pyrenaicum**, *Willd.*?

Acorus verus Bastard Calamus. 56. 2. **Acorus Calamus**, *L.*

"It prospereth exceeding well in my garden, but as yet it beareth neither flowers nor stalke." *Ger.* 57.

Adianthum nigrum Blacke Maiden haire. 975, 1. **Asplenium Adiantum-nigrum**, *L.*

Aethiopis Aethiopian mulleine. 634. **Salvia Æthiopis**, *L.*

Agrioriganum Field Organie. 541, 4. **Origanum vulgare**, *L.*

Album olus Corne Sallade. 242, 1. **Valerianella olitoria**, *Mœnch.*

Alcea arborea Tree Mallow. 788, 2. **Hibiscus syriacus**, *L.*

A. Veneta Venice Mallow. 791, 1. **H. Trionum**, *L.*

["... prospereth well in my garden from yeere to yeere." *Ger.* 792.

A. fruticosa pentaphyllea. 789, *first par.* **Althæa cannabina**, *L.*

Alchimilla Ladies mantle. 802. **Alchemilla vulgaris**, *L.*

Allium flore luteo Garlicke with yellow flowers. ——— **Allium Moly**, *L.*

A. vrsinum latifolium Ramsons. 141, 2. **A. ursinum**, *L.*

A. vrsinum angustifolium Ramsons with narrow leaves. ——— **A. ursinum**, *L. var.*

A. proliferum Harts Garlicke. 141, 1. *Ger. em.* 179, 1. **A. vineale**, *L.*

Alliaria Sauce alone or Iacke by the hedge. 650. **Erysimum Alliaria,** *L.*

Allyson Dioscoridis Dioscorides his Moonewoort. 379, *last par.* **Farsetia clypeata.** *R. Br.*

Alnus nigra Blacke Aller. 1286. **Rhamnus Frangula,** *L.*

Aloe Live long, or Aie greene. 409, *desc. not the fig.* **Aloe vulgaris,** *Lam.*

Alopecuros Foxetaile. 81. **Lagurus ovatus,** *L.*

Alleluia Wood Sorell. 1030, 1. **Oxalis Acetosella,** *L.*

Alsine repens Creeping Chickweede. ——— **Cucubalus bacciferus,** *L.*

A. folijs Trissaginis Chickweede with leaues like Germander. 492. 1. **Veronica agrestis,** *L.* (including **V. polita,** *Fr.*)

Althea arborea Hollihocke. 782, 4, *figs.* 1, ——— 5. **Althæa rosea,** *Cav.*

A. olbiæ French Mallow. 788, 1. **Lavatera Olbia,** *L.*

Lobel has written "Galloprovinciæ" in place of "French." "—— at the impression heerof, I have sowen some seedes of them in my garden, expecting the successe." *Ger.* 789.

A. floribus luteis Marsh Mallow with yellow flowers. 790. **Sida Abutilon,** *L.*

"The seede heerof is brought vnto vs from Spaine and Italy, we do yeerely sowe it in our gardens, the which seldome or never doth bring his seede to ripenes; by reason whereof, we are to seeke for newe seedes against the next yeere." *Ger. l. c.*

Amara dulcis Wood Nightshade. 279. **Solanum Dulcamara,** *L.*

A. dulcis flore albo Wood Nightshade with white flowers. 279, *second par.* **S. Dulcamara,** *L. var.*

"The other sort with white flowers I found in a ditch side against the garden wall of the right honorable the Earle of Sussex his house in Bermonsey streete by London, as you go from the court which is full of trees, vnto a farme house neere thereunto." *Ger. l. c.*

Amaracus Marierome. 538, 1. **Origanum Majorana,** *L.*

2. *A. folijs flauescentibus* Marierome with yellow leaues. 538, 2. **Origanum Onites,** *L. var.*

Amaranthus purpureus Purple Flower gentle. 254, 1. **Celosia cristata,** *L.*

A. maior Great Flower gentle. 254, 4. **Amaranthus caudatus,** *L.*

A. tricolor Variable Flower gentle. 254, 3. **A. tricolor,** *L.*

Ammi vulgatius Bishopsweede. 881, 1. **Ammi majus,** *L.*

A. Cræticum Candie Bishops weede. 881, 2. **Cachrys sicula,** *L.*

Amygdalus arbor The Almond Tree. 1256. **Amygdalus persica,** *L.*

Amomum Plinij Bastard Ginnie Pepper. 289. **Solanum Pseudocapsicum,** *L.*

Ampeloprasson Garlicke Leeke. ——— **Allium Ampeloprasum,** *L.*

Anblatum Broome Rape. 1130, 2. **Orobanche major,** *L.*

Anagallis flore cœruleo Blew Pimpernell. 494, 2. **Anagallis cærulea,** *All.*

A. flore phœniceo Tawnie Pimpernell. 494, 1. **A. arvensis,** *L.*

A. flore luteo Yellow Pimpernell. 494, 3. **Lysimachia nemorum,** *L.*

Anagyris Beane Trefoile. 1239, 1. **Anagyris fœtida,** *L.*

Anchusa Alkanet, or Red roote. 656, 1. **Alkanna tinctoria,** *Tausch.*

A. Neapolitana Naples Alkanet. 656, 2. **Onosma echioides,** *L.*

Androsaces Matthioli Sommers Nauell woort. 425, 2. **Androsace maxima,** *L.*

Androsemum Tutsan, or Parke leaues. 435. **Hypericum Androsæmum,** *L.*

Anemone maxima polyanthos The great double Windflower of Bithynia. 303, 3. **Anemone coronaria,** *L. var.*

A. coccinea multiplex Double Scarlet Windflower. 302, 2. **A. coronaria,** *L. var.*

A. coccineo simplici flore Single Scarlet Windflower. 303, 4. **A. coronaria,** *L.*

A. Gerani-folia Storks bill Windflower. 304, 7. **A. apennina,** *L.*

A. tuberosa radice Purple Windflower. 302, 1. **A. coronaria,** *L.*

A. tenuifolia simplici flore Small cut Windflower. 305, 10. **A. hortensis,** *L.*

A. tenuifolia flore cæruleo Blew Windflower. 303, 5. **A. hortensis,** *L.*

A. tenuifolia flore albo syluestris Small cut white Windflower. 305, 9. **A. trifolia,** *L.*

A. flore albo multiplex The double wilde white Windflower. 304, 8. **A. sylvestris,** *L.*

A. flore luteo Yellow Windflower. 306, 1. **A. ranunculoides,** *L.*

A. syluestris Wilde Windflowers of diuers colours. 306, 2. 307, 3, 4. **A. nemorosa,** *L.*

Angelica satiua Garden Angelica. 846, 1. **Archangelica officinalis,** *Hoffm.*

A. syluestris Wilde Angelica. 846, 2. **Angelica sylvestris,** *L.*

Anisum Anniseede. 880. **Pimpinella Anisum,** *L.*

" I haue often sowen it in my garden, where it hath brought foorth his ripe seede, when the yeere hath fallen out to be temperate." *Ger. l. c.*

Anteuphorbium Counterpoison Gumme thistle. 1014, 2. **Kleinia Anteuphorbium,** *DC.*

Anthemis duplici flore Double Cammomill. 616, 3. **Anthemis nobilis,** *L. var.*

Anthora Holsome Wolfesbane. 820. **Aconitum Anthora,** *L.*

Antirrhinum album White Snapdragon. 438,2. } **Antirrhinum majus,** *L.*

A. purpureum Purple Snapdragon. 438, 1. }

Anthyllis leguminosa Kidney Vetch. 1060, 1. **Anthyllis Vulneraria,** *L.*

Apios Fuchsij Knobbie Spurge. 407, 18. **Euphorbia Apios,** *L.*

Apocynum rectum Upright Dogs bane. 755, *first par.* **Marsdenia erecta,** *R. Br.*

A. repens Climing Dogs bane. 754. **Periploca græca,** *L.*

" My louing friend *John Robin* Herbarist in Paris, did sende me plants of both the kindes for my garden, where they flower and flourish." *Ger.* 755.

Aquilegiæ variæ Diuers sorts and colours of Colombines. 935, 1, 2. 936, 3. **Aquilegia vulgaris,** *L.*

Arabis Candie Mustard. 207. **Iberis umbellata,** *L.*

" This plant groweth naturally in that Pannonia which is nowe called Austria, in vntoiled places, and by high waie sides: in Crete or Candia, in Spaine and Italie, and such like hot regions, from whence I received seede, by the liberalitie of the right Honorable Lorde *Edwarde Zouche* at his returne into England from those partes, with many other rare seedes, which do flourish in my garden, for which I think myself much bounde vnto his good Lordship." *Ger. l. c.*

Aracus Wilde Vetch. 1052, 2. **Vicia hirsuta,** *Koch.* ?

Arbor Iudæ Iudas Tree. 1240. **Cercis Siliquastrum,** *L.*

Arbor Vitæ Tree of life. 1187. **Thuja occidentalis,** *L.*

Armeria syluestris Wilde Sweete Iohns. 478, 2. }

A. alba White Sweete Iohns. 478, 1. } **Dianthus Carthusianorum,** *L.*

A. rubramultiplex Double Red Iohns. 479. *third par.* }

A. polyanthos Double Sweete Williams. ——— **D. barbatus,** *L. var.*

A. alba guttata Spotted Sweete Iohns 479, *par.* 2. **D. superbus,** *L.*

A. suaue flore rubente Sweete Williams of a bright red. 479, 4. **D. barbatus,** *L.*

A. prolifera Sweete Williams, many in a hose. 479, 3 ? **D. barbatus,** *L. var.*

3. *Argemone Tabernæ-montani* Bastard Poppie. 300, 1. **Papaver hybridum,** *L.*

Arisarum latifolium Broad leafed Friers hood. 686, 1. **Arisarum vulgare,** *Targ.*

A. angustifolium Narrow leafed Friers hood. 686, 2. **Biarum tenuifolium,** *Sch.*

Aristolochia longa vera Long Birthwoort. 696, 1. **Aristolochia longa** *L.*

A. rotunda vera Round Birthwoort. 696, 2. **A. rotunda,** *L.*

A. clematitis Climing Birthwoort. 697, 3. **A. Clematitis,** *L.*

Artemisia leptophyllos Small iagged Mugwoort. 945, 2.
A. mater herbarum Mugwoort. 945, 1.
} **Artemisia vulgaris,** *L.*

A. marina Sea Mugwoort. 946, 3. **A. gallica,** *Willd.*

Asarina Italian Asarabacca. 688, 2. **Homogyne alpina,** *Cass.*

Asarum Common Asarabacca. 688, 1. **Asarum europæum,** *L.*

Asclepias White Swallowwoort. 751, 1. **Vincetoxicum officinale,** *Mœnch.*

A. flore nigro Blacke Swallowwoort. 751, 2. **V. nigrum,** *Mœnch.*

Ascyron Creticum S. Peter's woort of Candie. ——— **Hypericum hircinum,** *L.*

Asperula cærulea Blew Woodroofe. 965, *last par. but one.* **Asperula arvensis,** *L.*

Asphodelus albus White Asphodill. 86, 1. **Asphodelus albus,** *Willd.*

A. albus ramosus Branched Asphodill. 86, 2. **A. ramosus,** *L.*
The figures of the last two plants are transposed in the Herball.

A. bulbosus Bulbous Asphodill. 89. **Ornithogalum pyrenaicum,** *L.*
"It groweth in the gardens of herbarists in London, and not elsewhere that I know of: for it is not very common." *Ger. l. c.*

A. fistulosus Hollow Asphodill. 44, 3. *descr. not the fig. Ger. em.* 48, 3. **Asphodelus fistulosus,** *L.*

A. luteus Yellow Asphodill. 87, 4. **A. luteus,** *L.*

A. Lancastrensis Lancashire Asphodill. 88, 2. **Narthecium ossifragum,** *Huds.*
By an error Gerard employs this figure also for "*A. fistulosus.*" 44, 3.

Asparagus Sperage. 953, 1. **Asparagus officinalis,** *L.*

Asplenium Spleenewoort. 978, 1. **Ceterach officinarum,** *Desv.*

Aster Atticus flore luteo Yellow Starwoort. 392, 1. **Pallenis spinosa,** *Cass.*

A, Atticus flore cæruleo Blew Starwoort. 391, 4.? *no fig.* **Aster Amellus,** *L.*

A. inguinalis Flanke Starwoort. ——— **A. sp. ?**
This name is not given in *Bauh. Pin.* Probably one of the plants mentioned by Johnson (*Ger em.* 489) as in cultivation by Mr. Tradescant and others; a Michaelmas daisy, perhaps *A. Novi-Belgiæ, L.*

Astragalus Milke Vetch. 1058, 1. **Phaca bœtica,** *L.*

Astragaloides Little Wilde Milke Vetch. 1059, 4, *no fig.* **Astragalus hypoglottis,** *L.*

Astrantia Masterwoorts. 848, 1. **Imperatoria Ostruthium,** *L,*

A. nigra Blacke Masterwoorts. 828. **Astrantia major,** *L.*

Asureus conuoluulus Blew Bindweed. 715. **Pharbitis Nil,** *Chois.*
"The seede of this rare plant was first brought from Syria and other remote parts of the world, and is a stranger in these northren parts, yet have I brought vp and nourished it in my garden vnto flowring, but the whole plant perished before it could perfect his seed." *Ger. l. c.*

Atriplex hortensis alba White Garden Arach. 256, 1.
A. hortensis rubra Red Arach of the Garden. 256, 2.
} **Atriplex hortensis,** *L.*

A. olida Stinking Arach. 258. **Chenopodium Vulvaria,** *L.*

Atractylis Wilde Bastard Saffron. 1008, 1. **Kentrophyllum lanatum,** *D.C.*

A. hirsutior Hairie Bastard Saffron. 1008, 2. **Carduus benedictus,** *Gaert.*

Auricula vrsi flore purpureo Purple Beares eares. 640, 2.
A. vrsi flore luteo Yellow Beares eares. 640, 1.
A. vrsi flore vario Variable coloured Beares eares. ——
} **Primula Auricula,** *L.*

Auena nuda Naked Barley. 68, 2. **Avena nuda,** *L.*

B.

Bamia Strange Marsh Mallow. 787, 2. **Hibiscus palustris,** *L.*

Balsamina cucumerina Female Balsam apple. 290, 1. **Momordica Balsamina,** *L.*

B. fœmina Male Balsam apple. 290, 2. **Impatiens Balsamina,** *L.*

The English names of these plants are transposed, as noticed by Lobel. *MS. in loc.*

Balsamita mas Costmarie. 523. 1. **Pyrethrum Balsamita,** *L.*

B. fœmina Maudleine. 523, 2. **Achillea Ageratum,** *L.*

Barbarea S. Barbaraes woorts, or Winter Cresses. 188. **Barbarea vulgaris,** *R. Br.*

Barba capri Fuchsij Meadsweete. 886. **Spiræa Ulmaria,** *L.*

Bellides variæ Diuers sorts of Daisies. 509, *etc.* **Bellis perennis,** *L.*

Bellis prolifera Daisies, many on a stalke. ——— **B. perennis,** *L, var.*

Behen album Spatling Poppie. 550, 2. **Silene inflata,** *Sm.*

B. rubrum Catchflie. 481, 2. **S. Muscipula,** *L.*

4. *Berberis maximo fructu* Great Berberries. 1144. **Berberis vulgaris,** *L.*

B. sine acinis Berberies without stones. 1144, *last par. of descr.* **B. vulgaris,** *L. var.*

Beta rubra Red Beete. 251, 3. *no fig.* **Beta vulgaris,** *L. var.*

> "There is likewise another sort heerof that was brought vnto me from beyonde the seas by that courteous merchant master *Lete* before remembred, the which hath leaues very great and red of colour, as is all the rest of the plant, as well roote as stalkes and flowers, full of a perfect purple iuice tending to rednesse : - -. It grew with me 1596. to the height of viij. cubites, and did bring foorth his rough and vneeuen seede very plentifully : with which plant nature doth seeme to plaie and sport hirselfe : for the seedes taken from that plant which was altogither of one colour and sowen, doth bring foorth plants of many and variable colours, as the worshipful gentlemen master *Iohn Norden* can very well testifie, vnto whom I gaue some of the seedes aforesaide, which in his garden brought foorth many other of beautifull colours." *Ger.* 251,—2.

B. nigra Blacke Beete. ———
B. alba White, or Yellow Beete, 251, 1. } Varieties of **Beta vulgaris,** *L.*
B. Romana Romane Beete. 252, D.

Betonica flore albo Betonie with white flowers. 577, 2. **Stachys Betonica,** *Benth. var.*

Bistorta maior Snake Weede. 322, 1. **Polygonum Bistorta,** *L.*

B. maior altera Great Snake Weede. 322, 3. **P. Bistorta,** *L. var.*

B. minor Small Snake Weede. 322, 2. **P. viviparum,** *L.*

Blattaria flore luteo Yellow Moth Mulleine. 633, 1. **Verbascum Blattaria,** *L.*

B. flore purpureo Purple Moth Mulleine. 633, 2. **V. phœniceum,** *L.*

B. flore albo White Moth Mulleine. ——— **V. Blattaria,** *L. var.*

B. flore rubente Red Moth Mulleine. 633, *last par.* **V. phœniceum,** *L. var.*

Blitum album White Blites. 253, *descr. on* 252, *first par.*
B. rubrum Red Blites. 252, 4, *no fig.* } **Atriplex hortensis,** *L.*
B. supinum Flat Blites. 252, 2, *no fig.*

Bolbocastonon Earth Nuts. 906, 1. **Bunium flexuosum,** *Brot.*

Bonus Henricus False Mercurie. 259. **Chenopodium Bonus Henricus,** *L.*

Borago semper virens Euerliuing Borage. 653, 3. **Anchusa sempervirens,** *L.*

Botrys Oke of Jerusalem. 950, 1. **Chenopodium, Botrys,** *L.*

Brassica florida Coley flowers. 246, 9. ——
B. fimbriata Welted Cole woorts. —— } Varieties of **Brassica oleracea**, *L.*
B. tophosa Swolne Colewoorts. 246, 10.

"The swolen Colewoort of al other is the strangest, which I receued fro a worshipfull marchant of London master *Nicholas Lete*, who brought the seeds thereof out of Fraunce." *Ger. l. c.*

B. tricolor Variable coloured Colewoorts. ——
B. caulirapa Turnep Cole, or Rape Cole. ——
B. patula Open Cabbage Cole. 245, 6. } Varieties of **B. oleracea**, *L.*
B. arborescens Tree Colewoort. ——
B. exotica Sauoy Cole. 247, 11 & 12.

B. marina monospermos Sea Colewoorts. 248, 16. **Crambe maritima**, *L.*

Bugula flore albo White Bugle. 506, 2. **Ajuga reptans**, *L. var.*

Bulbus eriophorus Woolie Iacint. 106. **Scilla hyacinthoides**, *L.*

"Myselfe haue been possessed with this plant at the least 12. yeers, whereof I haue yeerely great increase of new rootes, but I did neuer see any token of budding or flowring to this day." *Ger.* 106-7.

"This flowred in the Garden of Mr. Richard Barnesley at Lambeth, onely once in the moneth of May, in the yeare 1606, after hee had there preserved it a long time: but neither he nor any else in England that I know, but those that saw it at that time, euer saw it beare flower, either before or since." *Park. par.* 124.

Miller mentions this as a variety of his *S. eriophora*, stating that it multiplies rapidly by offsets, but never flowered during thirty years whilst under his care. *Mill. Gard. Dict.*

Buphthalmus verus Dod. Oxe eie. 607, 2. **Anacyclus radiatus**, *Lois.*

C.

Cachrys vera Herbe Franckincense, 858, "2." **Seseli Libanotis**, *Koch.*

Cakile Serapionis Sea Rocket. 192, 5, *no fig.* *Ger. em.* 248, 5. **Cakile maritima**, *Scop.*

Calamenta montana Mountaine Calamint. 556, 1. **Calamintha cretica**, *Benth.*

C. prestantior Sweete Calamint. 556, 2. **C. Nepeta**, *Link & Hoffm.*

Calendulæ variæ Diuers sorts of Marigoldes. 600, *etc.* **Calendula officinalis**, *L.*

Chamædrys Germander. 530, 1. **Teucrium Chamædrys**, *L.*

C. laciniatis folijs Iagged Germander. 421, 2. *Ger. em.* 525, 2. **T. Botrys**, *L.*

Campanula lactescens Steeple Belflower. 367, 2. **Campanula pyramidalis**, *L.*

C. persicifolia Peach leafe Belflower. 367, 1. **C. persicifolia**, *L.*

C. persicifolia alba White Peach leafe Belflower. —— **C. persicifolia**, *L. var.*

C. elegantissima ex China Blew Belflower of China. —— May possibly be **Platycodon grandiflora**, *A. DC.* but there is no contemporaneous notice of this plant.

Canna Indica Indian Reed. 36. **Canna indica**, *L.*

"My selfe have planted it in my garden diuers times, but it neuer came to flowring or seeding." *Ger. l. c.*

Capnos fabacea radice Small Holewoorts. —— **Corydalis bulbosa**, *DC.*

C. alba White Holewoorts. —— **C. capnoides**, *Pers.*

Capparis vera Capers. 748, 1 & 2. **Capparis spinosa**, *L.*

"— myself at the impression heerof, planted some seedes in the bricke wals of my garden, which as yet doe spring and growe greene, the successe I expect." *Ger.* 749.

C. leguminosa Corne Capers. 750. **Zygophyllum Fabago**, *L.*

The seeds were sent from the "lowe countries" to Gerard. *v. Ger. l. c.*

Capsicum Actuarij Ginnie Pepper. 292, 1. **Capsicum annuum,** *L.*

C. Indicum Indian Pepper, 292, 2, *descr. not the fig.* **C. frutescens,** *L.*

5. *Caput Gallinaceum Belgarum* Dutch Cocks Head. 1062, 1. **Onobrychis sativa,** *Lam.*

Cardiaca Mother Woort. 569. **Leonurus Cardiaca,** *L.*

C. spinosa Camerarij Thornie Mother Woort. 559, 4. **Moluccella spinosa,** *L.*

Carduus stellatus Starrie Thistle. 1003, 1. **Centaurea Calcitrapa,** *L.*

C. acaulis Thistle without Stalke. ——— *Ger. em.* 1158, 3. **Carduus acaulis,** *L.*

C. tomentosus Woollie Thistle. 990, *par.* 5 ? *Ger. em.* 1152, 6. **C. eriophorus,** *L.*

C. globosus Globe Thistle. 990, *fig.* **Echinops sphærocephalus,** *L.*

C. Chrysanthemus Golden Thistle. 993, 1. **Scolymus hispanicus,** *L.*

Caryophyllorum hortensium variæ in colore differ. Gilloflowers of diuers sorts and colours. 472. **Dianthus Caryophyllus,** *L.*

Caryophyllus flore luteo Orange tawnie Gilloflowers. 472, *no fig.* **D. Caryophyllus,** *L. var.*

"The which a worshipfull marchant of London Master *Nicholas Lete*, procured from Poland, and gaue me thereof for my garden, which before that time was neuer seene nor heard of in these countries." *Ger. l. c.*

Caryophyllata Alpina Auance of the mountaine. 842, 2. **Geum montanum,** *L.*

C. rotundifolia Round leafed Auance. ——— Probably **Saxifraga rotundifolia,** *L.*

Carum Carawaies. 879. **Carum Carui,** *L.*

Castanea Chestnut. 1253, 1. **Castanea vesca,** *Gaert.*

Catanance The Inchaunting Vetch. ——— *Ger. em.* 1249, 2. **Lathyrus Nissolia,** *L.*

Caucalis Hispanica Spanish Bastard Parsley. 868, *descr.* **Krubera leptophylla,** *Hoffm.*

"I have sowen [it] in my garden, but it perished before the seede was perfected." *Ger. l. c.*

C. cretensis Bastard Parsley of Candie. 893, *descr. only.* **Tordylium officinale,** *L.*

"Narbon in Fraunce from whence I had seedes, which prosper well in my garden." *Ger. l. c.*

Caucason Withering Molie. 145, 5. **Allium magicum,** *L. var.*

Cauda muris Mousetaile. 345, 4. **Myosurus minimus,** *L.*

Centaureum flore albo White Centorie. 437, 1. **Erythræa Centaurium,** *L. var.*

C. luteum Yellow Centorie. 437, 2. **Chlora perfoliata,** *L.*

C. magnum Great Centorie. 436. **Centaurea Centaurium,** *L.*

C. magnum flore luteo Great Centorie with Yellow flowers. ——— *Ger. em.* 546, 2. **C. alpina,** *L.*

Cerasa Anglica serotina The Common English Cherrie. 1319, 1.
C. Belgica Flaunders Cherrie. 1319, 2.
C. alba Hispanica Spanish Cherrie. 1320, 3.
C. racemosa Grape Cherrie. 1320, 6.
C. agriotta The sower French Cherrie. 1323, *text.*
C. serotina altera The late ripe Cherrie. 1320, 5.
C. Gasconica Gascoine Cherrie. 1320, 4.
C. cordata maiora Great hart Cherrie.
C. cordata minora The lesser hart Cherrie. } 1323, *first par.*
C. nigra maiora The great blacke Cherrie.
C. nigra minora The lesser blacke Cherrie. } 1323, 11.
C. duplici flore Double floured Cherrie. 1321, 8.
C. duplici flore altera Double flowred Cherrie bearing fruit. ———
C. cærulea The blewish Cherrie. ———

} Varieties of **Prunus Cerasus,** *L.*

C. avium racemosa Birds Cherries. 1322, 9, 10. **Prunus Avium.** *L.*

Ceratia siliqua S. Iohns Bread, or Locust. 1241. **Ceratonia Siliqua,** *L.*

Cerinthe Plinij Honie woort. 431, 2. **Cerinthe minor,** *L.*

C. maior Great Honie woort. 431, 1. **C. major,** *L.*

"I have them in my garden; the seedes whereof I receaued of the right honorable the Lord *Zouch*, my honorable good friend." *Ger.* 432.

Cereus Peruanus The Pine Torch. 1015, 3. **Cereus peruvianus,** *Haw.*

This plant was supplied to Gerard by William Martin, and others, from the coast of Barbary, but it was destroyed by cold weather. *v. Ger.* 1016.

Ceruicaria maior Great Throate woort. 364, 1. **Campanula Trachelium,** *L.*

C. minor Small Throatwoort. 364, 4. **C. glomerata,** *L.*

C. Giganteum Giants Throatwoort. 365, 5. **C. latifolia,** *L.*

Chamæficus Dwarfe Fig Tree. 1327, *descr.* **Ficus Carica,** *L. var.*

It fruited abundantly each year, *v. Ger. l. c.*

Chamæcerasus Alpigena Dwarfe Cherrie tree. 1113, 5. **Lonicera alpigena,** *L.*

Chamælea tricoccos Widow waile. 1215. **Cneorum tricoccos,** *L.*

Chamælea Dwarfe Bay tree. 1216. **Daphne Mezereum,** *L.*

C. alpina glauca, argenteave Mountaine Widow Waile. 1217, 1. **D. alpina,** *L.*

Chamælinum pusillum Dwarfe Flaxe. 447, 4. **Linum catharticum,** *L.*

Chamæmalus Paradise Apple. 1277, *descr.* **Pyrus Malus,** *L. var.*

Chamæmorus Knotberries, or dwarfe Mulberries. 1090, 4. **Rubus Chamæmorus,** *L.*

6. *Chamæpitys* Lowe Pine. 421, 1. **Ajuga Chamæpitys,** *L.*

C. Austriaca Lowe Pine of Austrich. 422, *descr.* **Dracocephalum austriacum,** *L.*

Chamænerium Willow herbe with flowers like the Rose Bay. 386, 4. *Ger. em.* 497, 7. **Epilobium angustifolium,** *L.*

Chamæiris flore rubello Dwarfe Flowerdeluce with reddish flowers. ———

C. lutea Yellow Dwarfe Flowerdeluce. ———

C. niuea Snowe white Dwarfe Flowerdeluce. ———

C. purpurea Purple Flowerdeluce. ———

C. variegata Changeable Flowerdeluce. ———

C. angustifolia Narrow leafed Dwarfe Flowerdeluce. 52, 4.

C. violacea Violet coloured Dwarfe Flowerdeluce. ———

C. latifolia Broad leafed Dwarfe Flowerdeluce. ———

C. variegata Clusij Yellow and Purple Dwarfe Flowerdeluce. ———

} Varieties of **Iris pumila,** *L.*

C. tenuifolia Thinne leafed Dwarfe Flowerdeluce. 52, 5. **Iris graminea,** *L.*

Christophoriana S. Christophers herbe. 829. **Actæa spicata,** *L.*

"I haue received plants thereof from *Robinus* of Paris for my garden, where they flourish." *Ger. l. c.*

Chrysanthemum proliferum Flower of the sunne, many on one stalke. 613, 4.? **Helianthus multiflorus,** *L.*

C. Peruuianum Great Flower of the sunne. 612, 1. **H. annuus,** *L.*

". . it hath risen vp to the height of fourteene foote in my garden, where one flower was in waight three pounde and two ounces, and crosse ouerthwart the flower by measure sixteene inches broade." *Ger. l. c.*

C. aruorum Pasture marigold. 605, 1. **Chrysanthemum segetum,** *L.*

Cineraria Sea Ragweede. 218, 2. *Ger. em.* 280, 4. **Cineraria maritima,** *L.*

Circæa Inchaunters Nightshade. 280. **Circæa lutetiana,** *L.*

Cirsium Inchaunters Thistle. —— *Ger. em.* 1181, 1. **Cirsium monspessulanum,** *All.*
Cistus mas Male Holly Rose. 1093, 1. **Cistus parviflorus,** *Lam.*
C. fœmina Female Holly Rose. 1094, 5. **C. salvifolius,** *L.*
C. humilis Dwarfe Holly Rose. 1098, 17. **Helianthemum salicifolium,** *Pers.*
Cytisus Maranthæ Shrub Trefoile. 1124, 7. **Medicago arborea,** *L.*
C. siliquosus Codded Shrub Trefoile. 1122, 3. **Cytisus sessilifolius,** *L.*
Clematis peregrina flore albo Virgins Bowre. 741, 1. **Clematis Flammula,** *L.*
C. peregrina flore rubro Red Ladies Bowre. 740, 2. **C. Viticella,** *L.*
C. peregrina flore cæruleo Blew Ladies Bowre. 740, 1. **C. Viticella,** *L. var.*
C. Boetica Winter Ladies Bowre. 739, 2. **C. cirrhosa,** *L.*
C. Pannonica Bush Ladies Bowre. 742, 4. **C. integrifolia,** *L.*
C. Daphnoides Great Peruinkle. 747, *descr. no fig.* **Vinca major,** *L.*
Clynopodium The herbe Masticke. 544, 1. **Thymus Mastichina,** *L.*
Climenum Italorum Tutsan. 435. **Hypericum Androsæmum,** *L.*
Cnicus satiuus Bastard Saffron. 1007. **Carthamus tinctorius,** *L.*
Cochlearia Britannica English Scuruie grasse. 324, 2. **Cochlearia anglica,** *L.*
C. Batauorum Dutch Scuruie grasse. 324, 1. **C. officinalis,** *L.*
Colchicum Anglicum album White Medow Saffron. 127, 2. **Colcichum autumnale,** *L. var.*
C. Pannonicum Hungarie Meade Saffron. 127, 3. (*two figs.*) **C. autumnale,** *L.*
C. luteum Yellow Meade Saffron. 129, 8. *Ger. em.* 159. *Clus. Hist.* i. 164. **Sternbergia colchiciflora,** *W. & K.*
C. ephemerum Deadly Meade Saffron. 127, 1? **Colchicum autumnale,** *L.*
Colus Iouis Iupiters Distaffe. 627. *descr. not the fig.*? **Salvia glutinosa,** *L.*
Colutea Bastard Sena. 1116, 1. **Colutea arborescens,** *L.*
C. minima Dwarfe Bastard Sena. 1118, 5. **Coronilla valentina,** *L.*
C. scorpioides Scorpion Bastard Sena. 1116, 2. **C. Emerus,** *L.*
Condrilla rara flore purpurante Sowthistle with purplish flowers. 225, 4. **Crupina vulgaris,** *Pers.*
C. flore cæruleo Sowthistle with blew flowers. 224, 1. **Lactuca perennis,** *L.*
Coniza maior Great Fleabane. 390, 1. **Inula dysenterica,** *L.*
Conizæ variæ Diuers sorts of Fleabane. —— **Inula, sp.?**
Consolida media vulnerariorum Great Field Daisie. 509. **Chrysanthemum Leucanthemum,** *L.*
C. segetum Corne Daisie. 605, 1. **C. segetum,** *L.*
Consolidæ regales variæ Diuers sorts and colours of Larks heele. 922-3. **Delphinium Consolida,** *L.*
Conuoluuli varij Diuers sorts of Bindweeds. 712. **Convolvulus sepium,** *L.* & **C. arvensis,** *L.*
Coriandrum Corianders. 859. **Coriandrum sativum,** *L.*

"Coriander - - doth come of itself from time to time in my garden, though I neuer sowed the same but once." *Ger. l. c.*

7. *Cornus mas* Male Cornell Tree. 1282. **Cornus mas,** *L.*
C. fructu albo Cornell tree with white fruit. —— **C. mas,** *L. var.*
C. fœmina The Dogberrie tree. 1283. **C. sanguinea,** *L.*
Coronopus Harts horne. 346, 1. **Plantago Coronopus,** *L.*
C. Ruellij Bucks horne. 346, 2. **Senebiera Coronopus,** *Poir.*
Corona Imperialis Crowne Imperiall. 153, 11. **Fritillaria imperialis,** *L.*

C. terræ Alehoue. 705. **Nepeta Glechoma,** *Benth.*
Cortusa Matthioli Beares eare Sanicle. 645, 3. **Cortusa Matthioli,** *L.*
Cotyledon Wall Pennie woort. 423, 1. **Cotyledon Umbilicus,** *L.*
Crassula maior Great Orpin. 416, 1. **Sedum Telephium,** *L. var.*
Crateogonon Cowe Wheate. 84, 1. **Melampyrum pratense,** *L.*
Corylus Tripolitanus maximus Great Filberd of Tripolis. 1251, 3. **Corylus Colurna,** *L.*
Crocus Anglicus Common Saffron. 123, 1 & 2. **Crocus sativus,** *L.*
C. montanus Mountaine Saffron. ——— **C. nudiflorus,** *L.*
C. vernus flore albo Saffron of the spring with white flowers. 126, 4, *no fig.* **C. vernus,** *L. var.*
C. vernus flore luteo Saffron of the spring with Yellow flowers. 126, 3. *no fig.* **C. luteus,** *L.* or **C. susianus,** *L.*

"That pleasant plant that bringeth foorth yellow flowers was sent vnto me from *Robinus* of Paris, that painful and most curious searcher of Simples." *Ger.* 126.

C. vernus flore violaceo Saffron of the spring with Violet flowers. 126, 5, *no fig.* **C. vernus,** *L.*
C. vernus flore vario Variable spring Saffron. 125, 1. **C. versicolor,** *Ker.*
Cruciata herba Crosse woort. 965. **Galium Cruciata,** *L.*
C. gentiana Crosse woort Gentian. 351, 3. **Gentiana Cruciata,** *L.*
Cucumer Asininus Wilde Cucumbers. 766. **Momordica Elaterium,** *L.*
Cucurbitæ variæ Diuers sorts of Gourds. 777, *etc.* Several species of **Cucurbita.**
Cuminum satiuum Cumin seede. 907. **Cuminum Cyminum,** *L.*
Cupressus Cypresse tree. 1185. **Cupressus sempervirens,** *L.*

"The figure of this tree we do want, and the rather suffer it so to passe, because it is wel knowen to most." *Ger. l. c.*

Cyanus maior Great Corne flower. 592, 1. **Centaurea montana,** *L.*
Cyani varia genera Diuers sorts of Corne flower. 592-5. **C. Cyanus,** *L.*
Cyclamen folio hederæ Sowbread with leaues like Iuie. 694, 2. **Cyclamen hederifolium,** *Willd.*
C. orbiculato folio Sowbread with round leaues. 694, 1. **C. Coum,** *Mill.*
Cynara Artichoke. 991, 1. **Cynara Scolymus,** *L.*
Cynocrambe Dogs mercurie. 263, 1. **Mercurialis perennis,** *L.*
Cynoglossum Hounds toong. 659. **Cynoglossum officinale,** *L.*
C. pusillum Dwarfe Hounds toong. 659, *descr. only.* **C. sylvaticum,** *Haenke.*
C. cræticum Hounds toong of Candie. 659, *descr.* **C. cheirifolium,** *L.*

D.

Dactylo prunum The Date Plum. 1308. **Celtis australis,** *L.*
Daucus cræticus Candie Carots. 874. **Athamanta cretensis,** *L.*
D. selinoides Parsley Carots. 868, *fig.* *Ger. em.* 1021, 1. **Orlaya grandiflora,** *Hoffm.*
Dens caninus Dogs tooth. 154, 13.
Dentaria maior Great Dogs tooth. 154, 14, *fig.* 155, 3, *descr.*
D. alabastritis Violet Dogs tooth. 155, 2, *descr. only.* } **Erythronium Dens-canis,** *L.*
D. Rondeletij Leade woort. 1069. **Plumbago europæa,** *L.*
Digitalis alba White Foxe gloues. 646, 2. **Digitalis purpurea,** *L. var.*
D. flore luteo Yellow Foxe gloues. 646, *last par.* **D. lutea,** *L.*
D. purpurea Purple Foxe gloues. 646, 1. **D. purpurea,** *L.*

D. ferruginea Iron coloured Foxe gloves. 647, *first par.* **D. ferruginea,** *L.*

Dictamnum cræticum Dittanie of Candie. 651, 1. **Origanum Dictamnus,** *L.*

"I haue sowen it in my garden, where it hath flowred and borne seede; but it perished by reason of the iniurie of our extraordinarie colde winter that then hapned." *Ger. l. c.*

Doronicum Romanum Leopards bane. 620, 1. **Doronicum Pardalianches,** *L.*

Draba vera Treacle mustard. 204, 1. **Thlaspi arvense,** *L.*

D. altera Bowyers mustard. 204, 4. **Lepidium ruderale,** *L,*

Draco herba Tarragon. 193. **Artemisia Dracunculus,** *L.*

Dryopteris Small leafed Ferne. 974, 2. **Polypodium Dryopteris,** *L.*

E.

8. *Ebulus* Wallwoort, or Danewoort. 1238. **Sambucus Ebulus,** *L.*

Elatine Sharpe pointed Fluellin. 501, 2. **Linaria Elatine,** *L.*

E. fœmina Female Fluellin. 501, 1. **L. spuria,** *L.*

Elaphoboscum verum Wilde Parsnep. 870, 2, *descr. not the fig.* *Ger. em.* 1025, 2. **Pastinaca sativa,** *L.*

Elleborine Wilde white Hellebore. 358, 1.? **Epipactis latifolia,** *Sw.*

Epymedium Barren woort. 389. **Epimedium alpinum,** *L.*

"This rare and strange plant was sent to me from the French Kings Herbarist, *Robinus*, dwelling in Paris at the signe of the blacke head, in the streete called *Du bout du Monde*, in English, The end of the world. This herbe I planted in my garden, and in the beginning of May it came foorth of the ground, with small, hard and wooddie crooked stalkes: whereupon grow rough and sharp pointed leaues, almost like *Alliaria*, that is to say Sauce alone, or Iacke by the hedge. *L'Obelius* and *Dodonæus* say, that the leaues are somewhat like Iuie, but in my iudgement they are rather like *Alliaria*, somewhat snipt about the edges, and turning themselues flat vpright, as a man turneth his hande vpwardes when he receiueth money. Vpon the same stalks come foorth smal flowers, consisting of fower leaues, whose outsides are purple, the edges on the inner side red, the bottome yellow, and the middle part of a bright red colour, and the whole flower somewhat hollow. This I have seene, although *Dodonæus* saith that it neuer beareth any flower at all. The cause may be, for that the countrie where he sawe the same doth not agree so well with the nature of the plant, as our soile of England doth. The roote is small, and creepeth almost vpon the vppermost face of the earth. It beareth his seede in very small cods like Saracens Consound, but shorter: which came not to ripenesse in my garden, by reason that it was dried away with the extreme and vnaccustomed heate of the sunne, which happened in the yeere 1590. since which time from yeere to yeere it bringeth seede to perfection. Further *Dioscorides* and *Plinie* do report, that it is without flower or seede." *Ger. l. c.*

Eringium marinum Sea Hollie. 999, 1. **Eryngium maritimum,** *L.*

E. mediterraneum Mediterranean Sea Hollie. 999, 2. **E. campestre,** *L.*

E. planum Mountaine Sea Hollie. 1001, 1. **E. planum,** *L.*

Eruca peregrina Strange Rocket. 375, 4. **Vesicaria sinuata,** *Poir.*

E. nasturtio cognata Rocket, cosen to Cresses. 192, 4, *descr. only.* **Vella annua,** *L.*

Esula maior Germanica Quacksaluers Turbith. 404, 11. **Euphorbia palustris,** *L.*

E. minor Bastard Spurge. 405, 12, *descr. not the fig.* *Ger. em.* 502, 15. **E. platyphylla,** *Koch.*

E. exigua Dwarfe Spurge. 404, 12, *fig. only.* *Ger. em.* 503, 17. **E. exigua,** *L.*

E. rotunda Round leafed Spurge. 406, 15. **E. Peplus,** *L.*

Eupatorium Auicennæ King Kunigundus herbe. 574. **Eupatorium cannabinum,** *L.*

Euonymos Theophrasti Spindle tree. 1284, 1. **Euonymus europæus,** *L.*

F.

Faba Græcorum Greeke Beanes. 1036, 2. **Vicia narbonensis,** *L.*

Fabæ variæ Diuers sorts of beanes. 1038, etc. **Phaseolus, sp.**

Ferula galbanifera Fenell Giant. 899, *second par.* **Ferula Ferulago,** *L.*

F. sagapenifera Fenell Giant, bringing the Gumme Sagapenum. 898. **F. communis,** *L.*

F. nigra Blacke Fenell Giant. ——— Probably a variety of **F. Ferulago,** *L.*

Ferulago Little Fenell Giant. 899, *par.* 3. Possibly **F. glauca,** *L.*

Ficus de Algarua Fig of that part of Spain called Algarua. 1327. **Ficus Carica,** *L.*

F. Indica Fig of India. 1329, 2 *figs.* **Opuntia vulgaris,** *Mill.*

" I have bestowed great paines and cost in keeping it from the iniurie of our cold climate. It groweth . . at Zante, . . from whence he [*i. e.* William Marshall, Gerard's servant] brought me diuers plants thereof in tubs of earth, very fresh and greene for my garden, where they flourish as the impression heerof." *Ger.* 1330.

Filix florida Osmund the waterman. 971, 2 *figs.* **Osmunda regalis,** *L.*

F. mas Male Ferne. 969, 1. **Lastræa Filix-mas,** *Presl.*

F. fœmina Female Ferne. 969, 2. **Pteris aquilina,** *L.*

Flammula Speare woort. 814, 1. **Ranunculus Lingua,** *L.*

Filipendula Dropwoort. 900. **Spiræa Filipendula,** *L.*

Flos Adonis Adonis flower. 310, 1. · **Adonis autumnalis,** *L.*

F. Africanus maior Great Affrican Marigold. 611, 4. **Tagetes erecta,** *L.*

F. Africanus minor The lesser Affrican Marigold. 611, 5. **T. patula,** *L.*

F. Africanus simplex The single French Marigold. ——— **T. patula,** *L. var.*

" They . . . grow every where almost in Africke of themselues from whence we first had them, and that was when *Charles* the first Emperour of Rome made a famous conquest of Tunis." *Ger.* 611.

F. solis Flower of the Sunne. 612, 1. **Helianthus annuus,** *L.*

Fœnum Burgundiacum Burgundie hay. 1020, 2, *desc. only.* *Ger. em.* 1189, 2. **Medicago sativa,** *L.*

Ferrum equinum Horseshoe. 1056, 3. **Hippocrepis unisiliquosa,** *L.*

Fragaria sterilis Barren Strawberrie. 845, *descr. not the fig.* **Potentilla Fragariastrum,** *L.*

F. rubra Red Strawberrie. 844, 1.
F. alba White Strawberrie. 844, 2.
F. subviridis Greenish Strawberrie. 844, *bottom par.*
} **Fragaria virginiana,** *Ehrh.*

Fraxinus bubula Quicken tree. 1290. **Pyrus Aucuparia,** *Gaert.*

Fraxinella Bastard Dittanie. 1065. **Dictamnus albus,** *L.*

F. altera Great Bastard Dittanie. 1065, *par.* 2. **D. Fraxinella,** *Pers.*

Frittillaria Checkerd Daffodill. 122. **Fritillaria Meleagris,** *L.*

" The curious and painfull Herbarist of Paris *IohnRobin*, hath sent me many plants thereof for my garden." *Ger. l. c.*

Fumaria alba White Fumiterre. 929, 6. **Corydalis claviculata,** *DC.*

F. lutea Yellow Fumiterre. 928, 4. **C. lutea,** *DC.*

F. latifolia Broad leafed Fumiterre. 929, 5. **C. claviculata,** *DC. var.*

G.

9. *Galega* Goates Rue. 1068. **Galega officinalis,** *L.*

Galeopsis Pannonica Hungarie Dead Nettle. 568, 4. **Lamium Orvala,** *L.*

Gallium album White Ladies Bedstraw. 967, 2. **Galium palustre,** *L.*

G. luteum Yellow Ladies Bedstraw. 967, 1. **G. verum,** *L.*

G. flore rubro Red Ladies Bedstraw. 967, 3, *descr.* 968. **G. purpureum,** *L.*

" — the seede whereof was sent me from Argentine, or Strawsborough [Strasburg], in Germany." *Ger.* 968.

Genista Hispanica Spanish Broome. 1131, 3. **Spartium junceum,** *Link.*

Genistella Greening Weede. 1134, 1. **Genista tinctoria,** *L.*

Gelseminum album White Gessemin. 745, 1. **Jasminum officinale,** *L,*

Gentiana maior Great Felwoort, or Baldmoney. 351, 1. **Gentiana lutea,** *L.*

" Master *Isaac de Laune*, a learned phisition, sent me plants for the increase of my garden." *Ger.* 352.

G. Anglica English Felwoort. 354, 1 ? **G. campestris,** *L.*

Johnson pointed out the confusion of different plants in Gerard, and rewrote the whole of one chapter. *v. Ger. em.* 436.

Gentianella Little Felwoort. ——— *Ger. em.* 436, 1. **G. acaulis,** *L.*

" — is to bee found in most of our choice Gardens. As with Mr. *Parkinson*, Master *Tradescant* and Master *Tuggye*, &c." *Ger. em.* 437.

Geranium batrachioides Crowefoote Cranes bill. 797, 1, *in part.* **Geranium sylvaticum,** *L.*

G. bulbosum Bulbous Cranes bill. 795. **G. tuberosum,** *L.*

G. creticum Candie Cranes bill. 798, 1. **Erodium gruinum,** *Willd.*

G. fuscum Black Cranes bill. 799, 1. **Geranium phæum,** *Lam.*

G. gruinum Storks bill. 801, *par.* 1. **G. sanguineum,** *L. var.*

G. malacoides Bastard Cranes bill. 798, 2. **Erodium malacoides,** *Willd.*

G. repens Creeping Cranes bill. 800, 3 ? **E. cicutarium,** *Willd.*

G. Robertianum Herbe Robert. 794. **Geranium Robertianum,** *L.*

G. flore albo Storks bill with white flowers. 797, 2. **G. sylvaticum,** *L. var.*

G. flore cæruleo The Grace of God. 797, *in part, descr.* 796. **G. pratense,** *L.*

G. columbinum Doues foote. 793. **G. molle,** *L.*

G. nondum descriptum Storks bill, not yet described. 801, *par.* 2. **G. lucidum,** *L.*

Sent from Jean Robin to Gerard. *v. l. c.*

G. moschatum Musked Storks bill. 796, **Erodium moschatum,** *Willd.*

Gingidium Spanish Toothpikes. 885, 2. **Ammi Visnaga,** *Lam.*

Gladiolus Narbonensis French Corne Gladen. 95, 1. **Gladiolus communis.** *L.*

Glastum Woade. 394. **Isatis tinctoria,** *L.*

Glaux Dioscoridis Milke Tare. 1061. **Astragalus Glaux,** *L.*

G. exigua Little Milke Tare. 1059, *descr. par.* 4. **A. hypoglottis,** *L.*

G. vulgaris Common Milke Tare. ——— **A. glycyphyllos,** *L.*

Glycyrrhiza siliquosa Common Licorice. 1119, 2. **Glycyrrhiza glabra,** *L.*

G. echinata Hedgehog Licorice. 1119, 1. **G. echinata,** *L.*

Gnaphalium montanum Mountaine Cudweede. 516, 4 & 5. **Antennaria dioica,** *Gaert.*

G. marinum Sea Cudweede. 516, 3. **Diotis maritima,** *L.*

G. Anglicum English Cudweede. 515, 1. **Gnaphalium sylvaticum,** *L.*

G. Americanum Cudweede of America. ——— *Ger. em.* 641, 8. **G. margaritaceum,** *L.*

Gotne Alpina, siue Chamæpsillium Dwarfe Fleabane. —— **Plantago Psyllium,** *L. var.*

Gramen Parnasi Parnassus Grasse. 691, 2. **Parnassia palustris,** *L.*

G. striatum album Ladies Laces. 24, 2. **Phalaris arundinacea,** *L. var.* **variegata.**

Gratiola Hedge Hyssope. 466, 1. **Gratiola officinalis,** *L.*

G. Anglica English Hedge Hyssope. 466, 2, *descr. Ger. em.* 581, 3. **Scutellaria minor,** *L.*

"I found it growing vpon the bog or marrish ground at the further end of Hampsteed heath, and vpon the same heath towards London, neere vnto the head of the springs that were digged for water to be conueied to London 1590. attempted by that carefull citizen sir *Iohn Hart* Knight, Lord Maior of the Citie of London: at which time my selfe was in his Lordships company and viewing for my pleasure the same goodly springs, I found the said plant, not heretofore remembred." *Ger. l. c.*

Guiacum Patiuinum Italian Pockwood. 1310. **Diospyros Lotus,** *L.*

"I planted in the garden at Barne Elmes neere London two trees; besides there groweth another in the garden of Master *Graie*, an Apothecarie of London, and in my garden likewise." *Ger. l. c.*

H.

Harmala Wilde Rue. 1072, 5. **Peganum Harmala,** *L.*

Halicacabum Winter Cherries. 271, 1. **Physalis Alkekengi,** *L.*

Halymus Bastard Sea Purslane. 420, *par.* 2. **Atriplex Halimus,** *L.*

Hedysarum Hatchet Vetch. 1056, 1. **Coronilla varia,** *L.*

H. clypeatum Buckler Hatchet Vetch. 1056, *first par. no fig. Ger. em.* 1235, 7. **Hedysarum coronarium,** *L.*

10. *H. Glycyrrhizatum* Licorice Hatchet Vetch. 1056, 2. **H. coronarium,** *L. var.*

Hedypnois Wilde Cicorie. 220, 1. **Cichorium Intybus,** *L.*

Helleborastrum vtrunque Two sorts of wilde blacke Hellebore. —— Probably **Helleborus viridis,** *L.*

Helleborine radice repente Creeping wilde white Hellebore. 358, 1? **Epipactis latifolia,** *Sw.*

Helleborus niger verus True blacke Hellebore. 825, 1. **Helleborus niger,** *L.*

H. niger alter Setterwoort. 826, 3 & 4. **H. fœtidus,** *L.*

H. niger ferulaceus Oxe eie. 607, 1. **Adonis vernalis,** *L.*

H. albus White Hellebore. 356, 1. **Veratrum album,** *L.*

H. albus atrorubens White Hellebore, with flowers of a darke red colour. 356, 2. **V. nigrum,** *L.*

Helenium Elecampane. 649. **Inula Helenium,** *L.*

Helxine Pellitorie of the wall. 261. **Parietaria officinalis,** *L.*

H. cissampelos Blacke Bindeweede. 713, 4. **Polygonum Convolvulus,** *L.*

Hemerocallis Valentina Sea Onions of Valentia. 136, 2. **Pancratium maritimum,** *L.*

"I haue had plants of them brought me from sundry parts of the Mediterrane sea side, as also from Constantinople." *Ger.* 137.

Hemionitis sterilis Barren Spleenewoort. 977, *par.* 3. A small form of **Scolopendrium vulgare,** *L. v. Fl. Middx.* 341.

Hepatica nobilis flore albo Noble Liuerwoort with white flowers. 1031, *last par. no. fig.*
H. nobilis flore rubro Noble red liuerwoort. 1032, 2.
H. nobilis flore cæruleo Noble blew Liuerwoort. 1032, 1.
} **Hepatica triloba,** *Chaix. var.*

Herba Doria Captaine Doreas Woundwoort. 350. **Senecio Doria,** *L.*

H. Iudaica Glide woort. 565, 1, *descr. not the fig.* *Ger. em.* 700. **Lycopus europæus,** *L.*
H. Paris Paris herbe. 328, 1. **Paris quadrifolia,** *L.*
H. Turca Rupture woort. 454. **Herniaria glabra,** *L.*
H. Gerardi Goutwoort, or Herbe Gerard. 848, 2. **Ægopodium Podagraria,** *L.*
H. venti Rondeletij Windwoort. ——— *Ger. em.* 701, 2. **Phlomis herba-venti,** *L.*
Hermaphroditica orchis Butterflie satirion. 162, 1? **Habenaria bifolia,** *R. Br.*
Hermodactylus Italorum Veluet Flowerdeluce. 94, 2. **Iris tuberosa,** *L.*
Hieracium grandius Great Haukeweede. 232, 1. **Endoptera Dioscoridis,** *DC.*
Horminum verum Purple leafed Clarie. 628, 2. **Salvia Horminum,** *L.*
H. syluestre Wilde Clarie. 628, 1. **S. verbenaca,** *L.*
H. hortense Garden Clarie. 626, 1. **S. Sclarea,** *L.*
Hyacinthus Anglicus cæruleus English Blew Iacint. 99, 5.
H. Anglicus albus English White Iacint. 99, 6. } **Scilla nutans,** *Sm.*
H. Anglicus suaue rubens English reddish Iacint. 100, *par.* 2.
H. autumnalis Autumne Iacint. 98, 3. **S. autumnalis,** *L.*
H. botroides Blew Grape flower. 103, 3.
H. botroides albus White Grape flower. 104, 6. } **Muscari botryoides,** *Mill.*
H. botroides amœnus Skie coloured Grape flower. 104, 5.
H. orientalis cæruleus Blew Orientall Iacint. 100, 7.
H. orientalis albus White Orientall Iacint. 101, *par.* 3. } **Hyacinthus orientalis,** *L.*
H. orientalis Græcus Skie coloured Orientall Iacint. 100, 8.
H. orientalis brumalis Winter Iacint. 101, *par.* 4.
H. stellatus Fuchsij Starrie Iacint. 97, 1. **Scilla bifolia,** *L.*
H. stellatus Byzantinus Starrie Iacint of Turkie. 98, *par.* 4.? **S. amœna,** *L.*
H. stellatus Germanicus Starrie Iacint of Germanie. 97, 2. **S. Lilio-hyacinthus,** *L.*
H. comosus maior The greater faire haired Iacint. 103, 1.
H. comosus minor The lesser faire haired Iacint. ———
H. comosus Byzantinus Faire haired Iacint of Turkie. 102, *last par.* } **Muscari comosus,** *Mill.*
H. comosus albus Faire haired Iacint with white flowers. 103, 2.
Hyosciamus albus White Henbane. 283, 2. **Hyoscyamus albus,** *L.*
H. niger Blacke Henbane. 283, 1. **H. niger,** *L.*
H. luteus Yellow Henbane. 284. **Nicotiana rustica,** *L.*
Hypecoon Clusij Horned wilde Cumin. 909, 3. **Hypecoum procumbens,** *L.*
Hyppoglossum Bonifacia Horse Toong, or Double Toong. 761, 1. **Ruscus Hypoglossum,** *L.*
Hyssopus flore albo White flowred Hyssope. 465, 3. **Hyssopus officinalis,** *L. var.*
11. *H. tenuifolius* Iagged, or thinne leafed Hyssope. 465, 4.
H. latifolius Broad leafed Hyssope. 464, *last par.*
H. crispus Curlde Hyssope. 464, *par.* 4. } Probably garden varieties of **H. officinalis,** *L.*
H. Cræticus Hyssope of Candie. ———
H. niueus Anglicus English white Hyssope. ———
H. folijs flauescentibus Yellow leafed Hyssope. ———
Johnson substituted different figures for all of Gerard's. *v. Ger. em.* 579, *etc.*
Hypolapathum rotundifolium Bastard Rubarbe. 313, 6. **Rumex alpinus,** *L.*

I.

Iacea maior flore purpureo Great purple Knapweede, or Matfelon. 588, 2. **Centaurea Scabiosa**, *L.*

I. maior flore luteo Great yellow Knapweede, or Matfelon. 589, 3. **C. collina**, *L.*

I. maior flore flauo altera Another sort of great Knapweede. ——— Probably **C. solstitialis**, *L.*

Illecebra Wall Pepper. 415. **Sedum acre**, *L.*

Iris biflora Lusitanica Portingale Flowerdeluce. 49, 5. **Iris subbiflora**, *Brot.*

I. Florentina Orrice, or the Florentine Flowerdeluce. 47, 1. **I. florentina**, *L.*

I. Dalmatica maior pallida & cærulea Two sorts of the great Flowerdeluce of Dalmatia. 48, 3. **I. pallida**, *Lam.*

I. Dalmatica minor The little Dalmatian Flowerdeluce. 48, 4. **I. pallida**, *Lam. var.*

I. syluestris Bizantina peramœna Wilde Turkie Flowerdeluce. 52, 3. **I. sibirica**, *L.*

I. maritima Narbonensis The French marsh Flowerdeluce. 51, 3. *descr.* ? **I. graminea**, *L.*

I. Narbonensis minor The little French Flowerdeluce. 52, 4 ? **I. spuria**, *L.*

I. variegata Clusij Variable coloured Flowerdeluce. 51, 1. **I. variegata**, *L.*

I. violacea parua Little Violet Flowerdeluce. 49, 6. **I. pumila**, *L.*

I. Chalcedonica variegata The variable Flowerdeluce of Constantinople. 51, 2. **I. susiana**, *L.*

I. obsoleto flore Ouerworne Flowerdeluce. ———

I cannot find this mentioned in the *Herball*, but it may be the plant described by Parkinson, under the name of *Iris Purpura cærulea obsoleta labris fuscis* (*Park. Par.* 178.) which appears to be a form of **I. Xiphium**, *L.*

I. nostras palustris Common Waterflags. 46, 2. **I. Pseud-acorus**, *L.*

I. Susiana Blacke Flowerdeluce. 49, 8. **I. susiana**, *L.*

I. purpureo flore Purple Flowerdeluce. 46, 1. **I. germanica**, *L.*

I. bulbosa flore cæruleo Bulbose Flowerdeluce with blew flowers. 92, 1. **I. Xiphioides**, *Ehrh.*

I. bulbosa flore luteo Yellow bulbous Flowerdeluce. 93, 3. **I. lusitanica**, *Gawl.*

I. bulbosa flore vario Variable bulbous Flowerdeluce. 92, 2. } **I. Xiphium**, *L.*

I. bulbosa varia altera Another of greater beautie. ——— } **I. Xiphium**, *L.*

Iucca, [etc, the description as in Ed. 1. p. 9.] The roote whereof the bread Casaua or Cazaua is made. 1359. *Ger. em.* 1543. **Yucca gloriosa**, *L.*

Lobel altered a few words in the description, and added these at the end "sed temporæ tuberosior fit longeq: maior sese propagando. Hanc elapso [anno] descripsi, ex horto I. Gerardi Botanici Lond. plantarum auidissimi & amantissimi." *Lob. MS.* Thus the original description appears to have been drawn up by Lobel himself. *cf. Lob. Adv. alt. pars*, 507.

"This plant groweth in all the tract of the Indies, from the Magellane straights vnto the Cape of Florida, and in most of the Ilands of the Canibals, and others adioining, from whence I had that plant brought mee that doth growe in my garden, by a seruant of a learned and skilfull Apothecarie of Excester, named Master *Thomas Edwards*." *Ger. l. c.*

". . . our author . . . committed these errours: First, in that hee saith it is the root whereof Cazaua bread was made, when as *Lobel* in his description said he thought it to be *Alia species a Yucca Indica ex qua panis communis fit.* Secondly, in that he set downe the place out of the *Historia Lugd.* (who took it out of *Theuet*) endeauoring by that meanes to confound it with that there mentioned, when as he had his from Mr. *Edwards* his man. And thirdly (for which indeed he was most blameworthy, and wherein he most shewed his weaknesse) for that hee doth confound it with the *Manihot* or true *Yuicca*, . . . within some few yeares after our Author had set forth this worke it floured in his garden." *Ger. em. l. c.*

K.

Kali magnum Glassewoort, or Salt woort. 429. **Salicornia herbacea,** *L.*

K. minus Little Glassewoort, of some Frog grasse. ——— *Ger. em.* 535, 3. **Suæda maritima,** *Dum.*

It is not mentioned in *Ger.* but Johnson figures it, and gives Lobel's name as above.

Keyri multiplex varietas Diuers sorts of double Stocke Gilloflowers. 372, 2, *descr. not the fig. Ger. em.* 458, 2. **Matthiola incana,** *R. Br. var.*

Knawel siue Saxifraga altera Anglica Parsley piert. 453, *par.* 4, *not the fig.* **Alchemilla arvensis,** *Lam.*

L.

Lachryma Iobi Iobs Teares. 82. **Coix Lachryma,** *L.*

It ripened seed one year in his garden. *v. Ger. l. c.*

Lactucæ variæ Diuers sorts of Lactuse or Lettise. 239, *etc.* Varieties of **Lactuca sativa,** *L.*

Lactuca syluestris soporifera Sleeping wilde Lettise. ——— *Ger. em.* 309, 1. **L. virosa,** *L.*

Lagopus Hares foote. 1023, 2. **Trifolium arvense,** *L.*

L. maximus Great Hares foote. 1023, 1. *descr. in part, but not the fig. Ger. em.* 1192, 1. **T. incarnatum,** *L.*

Lamium album White Archangell. 567, 1. **Lamium album,** *L.*

12. *L. luteum* Yellow Archangell. 567, 2. **Galeobdolon luteum,** *Huds.*

L. Pannonicum Hungarie Dead Nettle. 568, 4. **Lamium Orvala,** *L.*

Lampsana Docke Cresses. 199, *descr. not the fig.* **Lapsana communis,** *L.*

Lanaria herba Mulleine. 629, 1. **Verbascum Thapsus,** *L.*

Lathyrus angustifolia Euerlasting Pease. 1053, *bottom par.* **Lathyrus sylvestris,** *L.*

L. latifolia Of the same with broader leaues. 1053, 1, *no fig.* **L. latifolia,** *L.*

Laurus Tynus Wilde Bay Tree. 1224, 1. **Viburnum Tinus,** *L.*

Lens Lentils. 1049, 1. **Ervum Lens,** *L.*

Lepidium Dittander, or Pepper woort. 187, 2. **Lepidium latifolium,** *L.*

Leucoium bulbosum precox maius Early bulbous Stocke Gilloflower. ——— **Leucojum vernum,** *L.*

L. bulbosum precox minus A lesser sort thereof. 120, 1. **Galanthus nivalis,** *L.*

L. bulbosum hexaphyllon Late flowring Sommer fooles. 120, 2. **Leucojum autumnale,** *L.*

L. triphyllon Early Sommer fooles, or Sommer sottekins. 121, 3. **L. æstivum,** *L.*

L. marinum Sea Stocke Gilloflowers. 374, 2? **Matthiola sinuata,** *R. Br.*

L. luteum multiplex Double Yellow Wall flowers. 371, 2. **Cheiranthus Cheiri,** *L. var.*

L. marinum creticum Candie Sea Stock Gilloflowers. ——— *Ger. em.* 459, 3. **Verbascum spinosum,** *L.*

Leuisticum Common Louage. 855, *descr. Ger. em.* 1008. **Levisticum officinale,** *Koch.*

L. verum True Louage. 892, *descr. Ger. em.* 1048, 1. **Laserpitium Siler,** *L.*

Johnson (*in loc.*) points out the confusion existing in Gerard's application of the figures.

Licium Italicum Boxe Thorne. 1151. **Rhamnus saxatilis,** *L.*

Lilium non bulbosum luteum Yellow Lillie. 90, 1. **Hemerocallis flava,** *L.*

L. non bulbosum phœniceum Orange Tawnie Lillie. 90, 2. **H. fulva,** *L.*

L. Alexandrinum Lillie of Alexandria. ——— **Ornithogalum arabicum,** *L.*

L. Bizantinum Lillie of Constantinople. 151, 9. **Lilium chalcedonicum,** *L.*

"This plant groweth wilde in the fields and mountaines, many daies iournies beyonde Constantinopole, whither it is brought by the poore pesants to be solde, for the decking vp of gardens. From thence it was sent among many other bulbs of rare & daintie flowers, by master *Harbran* ambassador there, vnto my honorable good Lord and master, the Lord Treasurer of England who bestowed them vpon me for my garden." *Ger. l. c.*

L. montanum Mountaine Lillie. 150, 7, 8. **L. Martagon,** *L.*

"The small sort I haue had many yeeres growing in my garden, but the greater I haue not had till of late, giuen me by my louing friend master *Iames Garret* apothecarie in London." *Ger.* 151.

L. rubrum Red Lillie. 148, 2.

This figure from Tabernæmontanus puzzled Bauhin, *Pinax*, 77, and Johnson omitted it, substituting in *Ger. em.* 192, 2, a figure of **L. bulbiferum,** *L.*

L. album White Lillie. 146, 1, **L. candidum,** *L.*

L. album Bizantinum White Lillie of Constantinople. 146, 2. **L. candidum,** *β, L.*

L. Persicum Persian Lillie. 152, 10. **Fritillaria persica,** *L.*

L. cruentum Blood Red Lillie. 149, 3. **Lilium bulbiferum,** *L.*

L. cruentum bulbiferum The bulbed Red Lillie. 149, *descr. par.* 2. *no fig.* **L. bulbiferum,** *ε, L.*

L. conuallium flore rubello May Lillie, or Conuall Lillie with red flowers. 331, 2. **Convallaria majalis,** *L. var.*

Limonium magnum Sea Lauender. 332, 1. **Statice Limonium,** *L.*

L. paruum Little Sea Lauender. 332, 2. **S. occidentalis,** *Lloyd.*

Linaria aurea Golden Toade flaxe. 442, 8. **Linosyris vulgaris,** *DC.*

L. Valentina Toade flaxe of Valentia. 441, 4. **Linaria supina,** *Desf.*

L. purpurea Purple Toade flaxe. 441, 3. **L. purpurea,** *Mill.*

Linum syluestre Wilde flaxe. 447, 2, *fig.* 3, *not descr.* **Linum angustifolium,** *Huds.*

L. marinum Sea flaxe. ——— *Ger. em.* 560, 7. **L. maritimum,** *L.*

Lotus tetragonolobus Square codded Pease. ——— *Ger. em.* 1198, 3. **Tetragonolobus purpureus,** *Mœnch.*

L. vrbanus Sweete Trefoile. 1025. **Melilotus cærulea,** *Lam.*

L. arbor Nettle Tree. 1308. **Celtis australis,** *L.*

"I haue a small tree thereof in my garden. There is likewise a tree thereof vnder London wall sometime belonging to *M. Gray*, an Apothecary of London; and an other great tree in a garden neere Colman streete in London, being the garden of the Queenes Apothecary at the impression hereof called M. *Hugh Morgan* a curious conserver of rare simples." *Ger. l. c.*

Lunaria, bolbonac White Sattin, or Honestie. 377, 1. **Lunaria biennis,** *Mœnch.*

L. raphanitis Sweete smelling White Sattin. 378, 2, *two figs.* **L. rediviva,** *L.*

L. minor Small Moone woort. 328. **Botrychium Lunaria,** *Sw.*

Lupinus satiuus Common Lupines. 1043, 1. **Lupinus albus,** *L.*

L. flore luteo Yellow Lupines. 1043, 2. **L. luteus,** *L.*

L. flore cæruleo Blew Lupines. 1043, 3. **L. varius,** *L.*

Lycopsis Wilde Bugloss. 658, 1. **Lycopsis arvensis,** *L.*

Lychnis agrestis multiflora alba Double field Campion. ——— **Lychnis vespertina,** *Sibth. var.*

L. agrestis multiflora rubra Double red Batchelers Buttons. ——— **L. diurna,** *Sibth. var.*

L. marina Anglica English Sea Campion. 382, 2. **Silene maritima,** *L.*

"by the sea side in Lancashire at a place called Lytham, fiue miles from Wygan, from whence I had some seedes brought me for my garden by a diligent searcher of simples, Master *Thomas Hesketh.*" *Ger.* 385.

L. coronaria alba White Campions. 381, 2. } **Lychnis coronaria,** *Desv. var.*
L. coronaria rubra multiplex Double red Campions. —— }

13. *L. chalcedonica* Campion of Constantinople, or None such. 380. **L. chalcedonica,** *L.*

Lylac Matthioli Blew Pipe. 1213, 2. **Syringa vulgaris,** *L.*

Lysimachia lutea Yellow Willow herbe. 386. **Lysimachia vulgaris,** *L.*

L. flore cæruleo Blew Willow herbe. 387, 5, *no fig. Ger. em.* 477, 9. **Veronica spicata,** *L.*

L. siliquosa Codded Willow herbe. 386, 3. **Epilobium hirsutum,** *L.*

L. spicata Spiked Willow herbe. 386, 2, *fig.* **Lythrum Salicaria,** *L.*

L. galericulata Hooded Willow herbe. 387, 6, *no fig.* **Scutellaria galericulata,** *L.*

" This I found in a waterie lane leading from the Lord Treasurers house called *Thibals* vnto the backside of his slaughter house." *Ger. l. c.*

Lythospermum maius Great Gromell. 486, 1. **Lithospermum purpureo-cæruleum,** *L.*

L. minus Little Gromell. 486, 2. **L. officinale,** *L.*

M.

Mala insana Mad or raging Apples. 274. **Solanum Melongena,** *L.*

" We haue had the same in our London gardens, where it hath borne flowers, but the winter approching before the time of ripening, it perished: notwithstanding it came to beare fruite of the bignes of a goose egge one extraordinarie temperate yeere, as I did see in the garden of a worshipfull merchant, Master *Haruie* in Limestreete, but neuer to the full ripenesse." *Ger. l. c.*

M. insana altera Yellow mad Apples. 274, *descr. only.* **S. Melongena,** *L. var.*

Mali persici decem varietates Ten sorts of Peaches. 1257, *etc.* Vars. of **Amydalus Persica,** *L.*

Malus arantia The Arange, or Orange tree. 1279, 3. **Citrus Aurantium,** *L.*

M. Armeniaca The Abrecocke, or Apricocke tree. 1260, 1 & 2. **Prunus Armeniaca,** *L.*

Malua Geranifolia Storks bill Mallow. 785, 4.? **Malva moschata,** *L.*

M. crispa Curled, or French Mallowes. 785, 3. **M. crispa,** *L.*

M. arborescens coccinei coloris Scarlet coloured Hollyhocke. 782, 1. }
M. arborea polyanthos rubro flore Double red Hollyhocks. 783, 4. } **Althæa rosea,** *L.*
Maluæ arboreæ variæ Diuers sorts of tree Mallowes. 783, 5. }

Malum punicum Pomegranate tree. 1262. **Punica Granatum,** *L.*

Marrubium album White Horehound. 561, 1. **Marrubium vulgare,** *L.*

M. creticum Candie Horehound. 562, 4. **M. peregrinum,** *L.*

Martagon imperiale Imperiall Lillie. 153, 11. **Fritillaria imperialis,** *L.*

Matricaria grato odore Sweete Feuerfew. 526, 1. **Pyrethrum Parthenium,** *Sm.*

M. duplici flore Double Feuerfew. 526, 2. **P. Parthenium,** *Sm. var.*

Medica Medicke Fodder. 1029, 1. **Medicago scutellata,** *Lam.*

M. spinosa Thornie Medicke Fodder. —— **M. intertexta,** *Willd.*

M. Arabica Medicke Fodder of Arabia. ——

Probably a form of the next species. *cf. R. Hist.* i. 963, 12.

M. Camerarij Germaine Medicke Fodder. 1021, 4. **M. maculata,** *Willd.*

M. marina Sea Medicke Fodder. 1029, 2. **M. marina,** *L.*

Melampyrum Cow Wheate. 84, 1. **Melampyrum pratense,** *L.*

Melanthium Damascenum Damaske Nigella. 925, 3. **Nigella damascena,** *L.*

M. flore luteo Yellow Nigella. —— **N. orientalis,** *L.*

M. flore albo White Nigella. —— **N. sativa,** *L.*

M. pleno flore Double Nigella. 925, 4. **N. sativa,** *L. var.*

Melilotus coronata Assyrian Clauer. 1033, 1. **Trigonella corniculata,** *L.*

M. Germanica vtraque Germaine Clauer of two sorts. 1034, 4. **Melilotus alba,** *Lam.* and 1034, 2, *descr.* **M. officinalis,** *L.*

M. Italica Italian Clauer. 1033, 2, *fig.* **M. italica,** *Lam.*

M. Arabica Clauer of Arabia. —— **Trigonella hamosa,** *L.* ?

Mentæ variæ Diuers sorts of Mints. 551, *etc.* **Mentha, spp.**

Melissa Common Balme. 558, 1. **Melissa officinalis,** *L.*

M. Turcica Turkie Balme. 558, 2. **Dracocephalum Moldavica,** *Lam.*

M. Moluca East Indian Balme. 559, 3. **Moluccella lævis,** *L.*

Melones saccharati varij Sugar and Muske Melons, diuers sorts. 771, 1 & 2. Varieties of **Cucumis Melo,** *L.*

"I haue seen at the Queenes house at Saint Iames very many of the first sort ripe, through the diligent and curious nourishing of them by a skilful Gentleman the keeper of the said house, called Master *Fovvle;* and in other places neere vnto the right Honorable, the Lord of *Sussex* house of Bermondsey by London, where from yeere to yeere there is very great plentie, especially if the weather be anything temperate." *Ger.* 772.

Melocoton The Melon Peach. —— **Amygdalus persica,** *L. var.*

Meon Spignell. 895. **Meum Athamanticum,** *Jacq.*

Mercurialis mas Male Mercurie. 262, 1. } **Mercurialis annua,** *L.*

M. fœmina Female Mercurie. 262, 2. }

Mespylus satiuus Common Medlar, or Open arse. 1265, 1. **Mespilus germanica,** *L.*

Morsus gallinæ Hen bit. 493, 4. **Lamium amplexicaule,** *L.*

M. gallinæ hæderaceus Iuie Hen bit. 493, 3. **Veronica hederifolia,** *L.*

14. *Mezereon* Dwarfe Bay. 1216. **Daphne Mezereum,** *L.*

Millefolium legitimum Common Yarrow, or Nose bleede. 914, 1. **Achillea Millefolium,** *L.*

M. rubrum Red Yarrow. 914, 2. A red flowered variety of **A. Millefolium,** *L.* from Holly Deane, near Sutton, Kent, *v. Ger. in loc.*

M. album White Yarrow. 915, 2 ? **A. nobilis,** *L.*

Millium Mill or Millet. 73, 1. **Panicum miliaceum,** *L.*

"I haue of it yeerely in my garden." *Ger.* 74.

M. Indicum Indian Millet. 75, 1, *etc.* **Zea Mays,** *L.*

Mirabilia Peruuiana The Maruell of Peru. 272. **Mirabilis Jalapa,** *L.*

" . . myselfe haue planted many yeeres, and haue in some temperate yeeres receiued both flowers and ripe seede." *Ger.* 273. (A long account of the plant is given, with directions for its preservation through the winter.)

Morus alba White Mulberrie. 1325, 2. **Morus alba,** *L.*

M. rubra Red, or purple Mulberrie.. 1325, 1. **M. nigra,** *L.*

Moluca spinosa Indian thornie Balme. 559, 4. **Moluccella spinosa,** *L.*

Moly Dioscorideum Moly, or Inchaunters roote. 143, 1. **Allium subhirsutum,** *L.*

M. Homericum Homers Moly. 144, 3. **A. magicum,** *L.*

M. Indicum Indian Moly. 144, 4. **A. magicum,** *L. var.*

M. serpentinum Serpents Moly. 143, 2. **A. multibulbosum,** *Jacq.* ?
M. folijs Narcissi Narcissus Moly. ——— **A. senescens,** *L.*
M. montanum latifolium Mountaine Moly. 142, 4. **A. Victorialis,** *L.*
Mollugo White Ladies Bedstraw. 967, 4. **Galium Mollugo,** *L.*
Monophyllon One blade. 330, 2. **Maianthemum bifolium,** *Lam.*
Morion Theophrasti Garden Nightshade. 281. **Mandragora vernalis,** *Bertol.*
Muscari flore luteo Yellow musked Grape flower. 105, 1. **M. macrocarpum,** *Sweet.*
M. cineritium Ash coloured Grape flower. 105, 2. **Muscari moschatum,** *Desf. var.*
Muscipula Catch flie. 481, 2. **Silene Muscipula,** *L.*
M. vera Birdlime woort. 481, 1. **Lychnis Viscaria,** *L.*
Mitulo Prunum, siue Prunum Mituli effigie The Muscle Plum. ——— **Prunus domestica,** *L. var.*
Myrrhida Plinij Mocke Cheruill. 796. **Erodium moschatum,** *Willd.*
Myrrhis Sweete Cheruill. 882, 2. **Myrrhis odorata,** *Scop.*
Myrtus Brabantica Gaule, or Sweete Willow. 1228. **Myrica Gale,** *L.*
Myrtacantha Butchers Broome. 759. **Ruscus aculeatus,** *L.*

N.

Narcissus luteus multiplex Yellow Daffodill double. 115, 2. **Narcissus Pseudo-narcissus,** *L. var.*
N. medio luteus Daffodill with the yellow circle. 110, 6. **N. biflorus,** *Curt.*
N. medio purpureus Purple circled Daffodill. 108, 1. } **N. poeticus,** *L.*
N. medio purpureus precox Early purple circled Daffodill. 108, 2. } **N. poeticus,** *L.*
N. medio purpureus precocior Timeliest purple circled Daffodill. 109, 3. } **N. poeticus,** *L.*
N. minor serotinus Late flowring little Daffodill. 110, 5. **N. serotinus,** *L.*
N. Pisanus Italian Daffodill, or Primerose peerelesse. 110, 8. **N. Tazetta,** *L.*
N. albus Bizantinus multiplex Turkie Daffodill. 111, 9. **N. orientalis,** *L.*

"The double white Daffodill of Constantinople was sent into England vnto the right Honorable the Lord Treasurer, among other bulbed flowers: whose rootes when they were planted in our London gardens, did bring foorth beautifull flowers, very white and double, with some yellowness mixed in the middle leaues, pleasant and sweete in smell; but since that time we neuer could by any industrie or manuring bring them vnto flowring againe." *Ger. l. c.*

N. albus Germanicus multiplex Double white Daffodill. ——— **N. poeticus,** *L. var.*
N. Persicus Persian Daffodill. 113, 13. **Sternbergia Clusiana,** *Ker.*
N. Iuncifolius Rush Daffodill. 112, 11. **Narcissus Jonquilla,** *L.*
N. totus luteus Single yellow Daffodill. 115, 2, *right hand portion* ? **N. incomparabilis,** *Curt.* ?
Nasturtium Indicum Indian Cresses. 196, *two figs.* **Tropæolum majus,** *L.*

"— receiued from my louing friend *Iohn Robin* of Paris." *Ger. l. c.*

N. crispum Curled Cresses. 194, *par.* 2, *no fig.* **Lepidium sativum,** *L. var.* ?

This was also sent by Robin to Gerard.

Nidus auis Birds Nest. 176, *descr. not the fig.* **Neottia Nidus-avis,** *Rich.*

Particular directions are given for finding the exact station near Gravesend, from which no doubt Gerard got the plant.

Nummularia Herbe Two pence. 505, 1. **Lysimachia Nummularia,** *L.*
Nux Iuglans Wall nut, or Walsh nut tree. 1252. **Juglans regia,** *L.*

N. vesicaria Bladder nut tree. 1249. **Staphylea pinnata,** *L.*

"It groweth . . . in the garden of the right Honorable the Lord Treasurer my very good Lord and Master, by his house in the Strand, . . also in my garden." *Ger. l. c.*

O.

Ocymum maximum Great Basill. 547, 1. **Ocimum Basilicum,** *L.*

O. minimum Bush Basil. 547, 3. **O. minimum,** *L.*

15. *Ocymoides* Cow Basill. 549, 3. **Mentha gentilis,** *L.*

Oenanthe aquatica Water Dropwoort. 902, 5. **Œnanthe fistulosa,** *L.*

O. cicutæ facie Hemlocke Dropwoort. 901, 4. **Œ. crocata,** *L.*

Oleander Rose Bay. 1220, 1. **Nerium Oleander,** *L,*

Oleaster Wilde Oliue. 1206, 2. **Olea europæa,** *L, a* **Oleaster,** *DC.*

Ononis flore albo Rest harrow with white flowers. 1141, 2. **Ononis spinosa,** *L. var.*

O. non spinosa Rest harrow without prickles. 1142, 3, *descr. not the fig.* **O. hircina,** *Jacq.*

Ophioglossum Adders toong. 327. **Ophioglossum vulgare,** *L.*

Ophioscorodon Mountaine Garlicke. 142, 3. **Allium Scorodoprasum,** *L.*

"I receiued a plant of it from M. *Tho. Edwards*, apothecarie in Excester, who found it growing in the west parts of England." *Ger.* 143.

Orchides variæ Diuers sorts of Satyrions, besides these folowing.

Orchis andrachnitis Maimed Satyrion. 166, 14, *descr. Ger. em.* 216, 14. **Ophrys aranifera,** *Huds.*

O. melittias Waspe Satyrion. 163, 5, *descr. not the fig. Ger. em.* 213, 5. **O. arachnites,** *Willd.* ?

O. ornithophora Birds Satyrion. 165, 9. **Habenaria bifolia,** *R. Br.*

O. apifera Humblebee Satyrion. 162, 3. *Ger. em.* 212, 3. **Ophrys apifera,** *Huds.*

O. spiralis Yellow Ladies traces. 167, 2. **Spiranthes autumnalis,** *Rich.*

O. radice repente Satyrion without stones. 175, 4. **Goodyera repens,** *R. Br.*

O. odorata Sweete stones. 167, 1. **Herminium Monorchis,** *R. Br.*

Ornithogalum Stars of Bethlem. 132, 1. **Ornithogalum umbellatum,** *L.*

O. Pannonicum Stars of Hungarie. 132, 4. **O. comosum,** *Willd.*

O. luteum Yellow Star of Bethlem. 132, 3. **Gagea lutea,** *Schult.*

Ornithopodium Birds foote. 1061, *par.* 1, *no fig.* **Ornithopus perpusillus,** *L.*

Origanum Cræticum Organie of Candie. 541, 3. **Origanum creticum,** *L.*

"The roote endured in my garden and the leaues also greene all this winter long, 1597. although it hath been saide that it doth perish at the first frost." *Ger.* 542

Orobus True Cich Pease. 1051. **Ervum Ervilia,** *L.*

Othonna polyanthos The great double Affrican Marigold. 609, 1. **Tagetes erecta,** *L. var.*

Oxalis rotundifolia Round leafed Sorrell. 320, 4. **Rumex scutatus,** *L.*

P.

Paliurus Christ his thorne. 1153. **Paliurus aculeatus,** *Lam.*

"I haue a small tree growing in my garden, . . by sowing of the seede." *Ger. l. c.*

Panax Chironium Chirons All-heale. 850, 1. **Opoponax Chironium,** *Koch.*

P. Asclepium Asclepiades his All-heale. ——— *Ger. em.* 1057, 3. **Ferula nodiflora,** *L. in part.* **F. sulcata,** *Desf.* ?

P. Heracleum Hercules his All-heale. 850, 2. **Heracleum Panaces,** *L.*

P. Mentastrifolium Clownes All-heale. 852. **Stachys palustris,** *L.*

Panicum album Italian Oatemeale, or white Panicke. 78, 1. **Panicum italicum,** *L.*

P. rubrum Americanum Red Panicke. —— *Ger. em.* 84, 3. **Pennisetum typhoideum,** *Rich.*

Papauer simplex purpureo flore Purple single Poppie. 299, 1. **Papaver Rhœas,** *L.*

P. simplex flore albo White single Poppie. 296, 1. **P. somniferum,** *L.*

P. polyanthos rubro flore Double red Poppie. 297, 5. **P. Rhœas,** *L. var.*

P. polyanthos albo flore Double white Poppie. 296, 4. **P. somniferum,** *L. var.*

P. corniculatum flore luteo Yellow horned Poppie. 294, 1. **Glaucium luteum,** *Scop.*

P. corniculatum violaceo flore Violet horned Poppie. 294, 3. **Rœmeria hybrida,** *DC.*

P. corniculatum phœniceo flore Red horned Poppie. 294, 2. **Glaucium corniculatum,** *Curt.*

Papus orbiculatus Bastard Potatoes. 781. **Solanum tuberosum,** *L.*

P. Hispanorum Spanish Potatoes. 780. **Batatas edulis,** *Chois.*

" I planted diuers rootes (that I bought at the exchange in London) in my garden, where they flourished vntill winter, at which time they perished and rotted." *Ger. l. c.*

Paronychia alsinæ folio Chickweede Naile woort. 499, 1. **Draba verna,** *L.*

P. rutaceo folio Rue Naile woort. 499, 3. **Saxifraga tridactylites,** *L.*

Parthenium Alpinum Mountaine Feuerfew. 527, *par.* 3, *no fig.* **Ptarmica atrata,** *DC.*

Pæonia mas Male Pionie. 830, 1. **Pæonia corallina,** *Retz.*

P. fœmina Female Pionie. 830, 2.

P. polyanthos Double Pionie. 831, 3.

P. promiscua Misbegotten Pionie. 830, *line* 1, *no fig.*

P. albicans Whitish Pionie. 831, *par.* 3, *no fig.*

} **P. officinalis,** *Retz.*

Pecten Veneris Venus Combe. 884. **Scandix Pecten-Veneris,** *L.*

Pentaphyllum maximum Great Cinquefoile. 836, 2. (1, *in descr.*) *Ger. em.* 987, 2. **Potentilla recta,** *L.*

16. *P. album* White Cinquefoile. 836, 3. (2 *in descr.*) *Ger. em.* 987, 3. **P. recta,** *L.? var.*

P. rubrum Red Cinquefoile. 836, 4. **P. Comarum,** *L.*

" . . . in a marrish ground adioining to the land called Bourne pondes, halfe a mile from Colchester; from whence I brought some plants for my garden." *Ger.* 839.

Peplis Hyssope Spurge. 406, 16. **Euphorbia Peplis,** *L.*

Peplios Round Spurge. 406, 15. **E. Peplus,** *L.*

Perfoliata Thorough Waxe. 430, 1. **Bupleurum rotundifolium,** *L.*

P. siliquosa Codded Thorough Waxe, 430, 2. **Erysimum orientale,** *R. Br.*

Periclymenum Woodbinde, or Honisuckles. 743, 1. **Lonicera Periclymenum,** *L.*

P. perfoliatum Double Honisuckles. 743, 2. **L. Caprifolium,** *L.*

P. arborescens Tree Honisuckles. 1111, 1. **L. Xylosteum,** *L.*

Periploca recta Vpright Dogs bane. 755, *descr. par.* 1. **Marsdenia erecta,** *R. Br.*

P. repens Climing Dogs bane. 754. **Periploca græca,** *L.*

Percipier Anglorum Parsley Breakstone. 453, 3. **Scleranthus annuus,** *L*

Petasites Butterburre. 668, 1 & 2. **Petasites vulgaris,** *Desf.*

Petroselinum Macedonicum verum Parsley of Macedonia. 864, 1, *fig. only.* **Athamanta macedonicum,** *Spr.*

P. crispum & complicatum Crispe, or curled Parsley. 861, 2. **Petroselinum sativum,** *L.*

Peucedanum Sulphur woort, or sea Fenell. 896, 1. **Peucedanum officinale,** *L.*

Phalangium ramosum Branched Spider woort. 44, 1. **Anthericum ramosum,** *L.*

P. non ramosum Spider woort without branches. 44, 2. **A. Liliago,** *L.*

Phalaris Alpisti, or Canarie seede. 80, 1. **Phalaris canariensis,** *L.*

Phaseoli varij Diuers sorts of French, or Kidney beanes. 1038, *etc.* **Phaseolus, spp.**

Phyllitis Harts toong. 976, 1. **Scolopendrium vulgare,** *Sm.*

P. multifido folio Finger Ferne, or branched Harts toong. 976, 2. **S. vulgare,** *Sm. var.*

Phyllirea Mocke Priuet. *Note.*—This probably includes two plants, *viz.* 1209, 1. **Phillyrea angustifolia,** *L.* and 1209, 2. **P. media,** *L.*

P. serratis folijs Iagged Mocke Priuet. 1210, 3. **P. latifolia,** *L.*

"These plants . . . I planted in the garden at Barne Elmes neere London, belonging to the right Honorable the Earle of Essex; I haue them growing in my garden likewise." *Ger.* 1210.

Pimpinella Burnet. 889, 1. **Poterium Sanguisorba,** *L.*

Pinguicula Butter woorts. 644, 2. **Pinguicula vulgaris,** *L.*

Pinus The Pine tree. 1173. **Pinus Pinea,** *L.*

Pinaster The wilde Pine tree. 1175, 1. **P. sylvestris,** *L.*

Pistacia The Pisticke, or Fisticke nut tree. 1248. **Pistacia vera,** *L.*

Pisum cordatum Hart Pease. 271, 2. **Cardiospermum Halicacabum,** *L.*

P. vmbelliferum Tufted, or Scottish Pease. 1045, 3.
P. excorticatum Pease without parchment in the cods. 1045, 4. } **Pisum sativum,** *L.*

P. minus ex luteo virescens Yellow flowring Pease. 1046, 5.
P. perenne Euerlasting Pease. 1046, 6. } **Vicia pisiformis,** *L.*

Plantago rosea Rose Plantaine. 340, 6. *Ger. em.* 420, 5. A monstrosity of **Plantago major,** *L.*

P. rosea incana Hoarie Rose Plantaine. 340, 5? *Ger. em.* 420, 4. A similar monstrosity of **P. media,** *L.*?

P. marina Sea Plantaine. 343, 3. **P. maritima,** *L*

Platanus verus The Plane tree. 1304. **Platanus orientalis,** *L.*

"My seruant, *William Marshall*, whom I sent into the Mediterranean sea, as chirurgion vnto the Hercules of London, found diuers trees heerof growing in Lepantæ, hard by the sea side, at the entrance into the towne, a port of Morea, being a part of Greece, and from thence brought one of those rough buttons, being the fruit thereof." *Ger. l. c.*

Polemonium Makebate, or shrub Trefoile. 1129. **Jasminum fruticans,** *L.*

Polium montanum Puliole mountaine. 528, 2. **Teucrium Polium,** *L.*

". . . by the gift of L'Obelius." *Ger.* 529.

Polygala flore albo White Milke woort. 449, 4.
P. flore cærulea Blew Milke woort. 449, 2.
P. rubens Red Milke woort. 449, 3. } **Polygala vulgaris,** *L.*

Polygonatum Salomons seale, or White roote. 756, 1. **Polygonatum multiflorum,** *All.*

P. Pannonicum Broad leafed Salomons seale. 756, 3. **P. officinale,** *All.*

"*Carolus Clusius* . . . sent [this] to London vnto Master *Garth* a worshipfull Gentleman, and one that greatly delighteth in strange plants, who very louingly imparted the same vnto me." *Ger.* 757.

P. minus Little Salomons seale. 756, 2. **P. verticillatum,** *All.*

Polygoni varia genera Diuers sorts of Knotgrasse. 453, *etc.*

Probably the two mentioned *in loc.* as growing in Gerard's garden were *Scleranthus annuus* and *Alchemilla arvensis* previously mentioned.

Polyspermum, Casani bassi Spotted Blites. 257, 3. **Chenopodium polyspermum,** *L.*

Poma amoris rubro fructu Red Apples of loue. 275. **Lycopersicum esculentum,** *Mill.*

P. amoris flaua Yellowish Apples of loue. 275, *par.* 2. **L. esculentum,** *Mill. var.*

P. Aegyptia The Aegyptian Apple. 276. **Solanum Æthiopicum,** *L.*

" . . . mine perished at the first approach of winter." *Ger. l. c.*

Pomum spinosum Thorne Apple. 277, 2. *fig. absent.* **Datura Stramonium,** *L.*

" . . . whose seedes I receiued of the right Honorable the Lord *Edward Zouch,* which he brought from Constantinople." *Ger.* 277.

The *Herball* was printed with a blank space where the figure of this plant should have been, in some copies, an impression from a block, which projects into the margin, is pasted over the blank space.

Populus alba White Poplar tree. 1301, 1. **Populus alba,** *L.*

17. *Poterion* Burnet Goates thorne. 1148, 3. **Poterium spinosum,** *L.*

" I have sowen the seede of Poterion in April which I receiued from *Ioachimus Camerarius* of Noremberge, that grew in my garden two yeres togither, and after perished by some mischance." *Ger. l. c.*

Primula veris flore rubro Birds eies, or Birdeine. 639, 1. **Primula farinosa,** *L.*

P. veris viridi flore Greene Primeroses. 637, 7. } Varieties of **P. vulgaris,** *Sm.*

P. veris viridi multiplici flore Greene Primeroses double. ——— }

P. veris flore geminato Cowslips, two in a hose. 636, 4. **P. veris,** *L. var.*

P. veris maxima Anglica Double Paigles. 636, 3. **P. vulgaris,** *Sm. var.*

Primulæ syluarum variæ Diuers sorts of field Primeroses. ———

Prunella flore albo Selfeheale with white flowers. 508, 3, *not the fig.* **Prunella grandiflora,** *L.*

Pruni arboris species triginta Plum trees, thirtie sorts. 1311, *etc.* **Prunus domestica,** *L.*

"' . . . my selfe haue three score sorts in my garden, and all strange and rare.'' *Ger.* 1311. " The greatest varietie of these rare plums are to be found in the grounds of Master *Vincent Pointer* of Twicknam." *Id.* 1313.

Pseudo-dictamnum Bastard Dittanie. 651, 2. **Ballota Pseudo-dictamnus,** *Benth.*

Pseudo-costus Wilde Valerian of the mountaine. 850, 1 ? **Opoponax Chironium,** *Koch.*

Psyllium Fleabane. 471, 1. **Plantago Psyllium,** *L.*

P. semper virens Euer greene Flebane. 471, 2. **P. Cynops,** *L.*

Pseudo-narcissus luteus multiplex Yellow Daffodill double. 115, 1. } **Narcissus Pseudo-**

Pseudo-narcissus Common Yellow Daffodill, 115, 2, *in part.* } **narcissus,** *L.*

P. Hispanicus maior Yellow Daffodill of Spaine. ——— **N. major,** *Curt.*

P. Hispanicus minor Little Spanish Daffodill. ——— **N. minor,** *L.*

Pseudo-bunium S. Barbaraes woort, or Winter Cresses. 188. **Barbarea vulgaris,** *R. Br.*

Ptarmica Sneeze woort, or Wilde Pellitorie. 483, 1. **Achillea Ptarmica,** *L.*

P. duplici flore Double Sneeze woort. 483, 2. **A. Ptarmica,** *L. var.*

Pulegium erectum Great Penniroyall. 545, 2. *Ger. em.* 671, 2. **Mentha Pulegium,** *L. var.* **[erecta.**

P. regale supinum Lesser Penniroyall. 545, 1. **M. Pulegium,** *L.*

Pulmonaria vera Cowslips of Jerusalem. 662, 1. **Pulmonaria officinalis,** *L.*

Pulsatilla Purple Passeflower. 308, 1. **Anemone Pulsatilla,** *L.*

Pyracantha Boxe thorne. 1151, 1. **Rhamnus saxatilis,** *L.*

Pyrethrum officinarum True Pellitorie of Spaine. 618, 1. **Anacylus Pyrethrum,** *Cass.*

Pyrola Winter greene. 330, 1. **Pyrola rotundifolia,** *L.*

Q

Quadrifolium phæum Purple woort, or blacke Three leafed grasse. 1028, 2. **Trifolium repens**, *L. var.*

Quinquenervia rosea Rose Ribwoort. 341, 2. **Plantago lanceolata**, *L. var.*

R.

Radix caua flore purpureo Purple Hollow roote. 931, 1.
R. caua flore albo White Hollow roote. 931, 2. } **Corydalis bulbosa**, *Pers.*

R. caua viridi flore Greene Hollow roote. 933, 10. **Adoxa moschatellina**, *L.*

Ranunculus Alpinus Crowfoote of the Alpes. 805, 4. **Ranunculus aconitifolius**, *L.*

R. magnus Anglicus polyanthos Double Yellow Crowfoote. 810, 1. **R. acris**, *L. var.*

R. bulbosus Bulbous Crowfoote. 806, 6. **R. bulbosus**, *L.*

R. Illyricus Crowfoote of Illyria. 806, 5. **R. illyricus**, *L.*

R. niueus polyanthos Double white Crowfoote. 812, 1. **R. aconitifolius**, *L. var.*

R. gramineus Grasse Crowfoote. 808, 10. **R. gramineus**, *L.*

R. globosus Globe Crowfoote, or Locker gowlons. 809, 13. **Trollius europæus**, *L.*

R. auricomus duplici flore Gold Thrum Crowfoote double. 810, 2. **Ranunculus auricomus**, *L. var.*

" . . . hath of late beene brought foorth of Lancashire vnto our London Gardens, by a curious gentleman in the serching foorth of Simples Master *Thomas Hesketh*, who found it growing wilde in the towne fields of a small village called Hesketh, not far from Latham in Lancashire." *Ger.* 810-11.

R. Tripolitanus Crowfoote of Tripolis in Syria, or red Crowfoote. 812, 2. **R. asiaticus**, *L. var.*

R. echinatus Hedgehog Crowfoote. 805, 3. **R. arvensis**, *L.*

Raphanus Biting Radish. 183, 1.
R. niger Blacke Radish. 183-4, 4.
R. pyriformis Round rooted blacke Radish. 184, 4. } Varieties of **Raphanus sativus**, *L.*

R. rusticanus Horse Radish. 187, 1. **Cochlearia Armoracia**, *L.*

Rhamnus Bucke thorne. 1154. **Rhamnus catharticus**, *L.*

Rhabarbarum monachorum Munks Rubarbe. 313, 6. **Rumex alpinus**, *L.*

Rheseda Plinij Crambling Rocket. 216, 1? **Reseda lutea**, *L.*

18. *R. maior* Great Crambling Rocket. 216, 2. **R. alba**, *L.*

Rhodia radix Rose woort, or Rose roote. 426. **Sedum Rhodiola**, *DC.*

" . . . Ingleborough Fels . . . from whence I have had plants for my garden." *Ger. l. c.*

Rhus siue Sumach Smacke, or Sumacke. 1291, 1. **Rhus Coriaria**, *L.*

Ribes nigra Blacke Corrans. —— *Ger. em.* 1593, 3, *no fig.* **Ribes nigrum**, *L.*

R. alba White Corrans. —— *Ger. em.* 1593, 2.
R. rubra Red Corrans. —— *Ger. em.* 1593, 1. } **R. rubrum**, *L.*

Ricinus The hand of God, or Palma Christi. 399, 1. **Ricinus communis**, *L.*

Rosa Anglica alba simplici flore The English white Rose single. —— **Rosa arvensis**, *Huds.*

R. Anglica alba multiplex The white Rose double. 1079, 1. **R. alba,** *L.*

R. rubra The Red Rose. 1079, 2. **R. gallica,** *L.*

R. rubra flore maximo The great red Rose, or red Prouince Rose. 1080, *last par.* ? Probably some variety of **R. centifolia,** *L.*

R. Damascena flore multiplici The great Holand Rose, comonly called the Prouince Rose. 1081, 6. **R. centifolia,** *L.*

R. prouincialis The common Damask Rose. 1079, 3. **R. provincialis,** *Ait.*

R. moschata simplici flore The single Muske Rose. 1084, 1.
R. moschata multiplex The double Muske Rose. 1084, 2.
R. moschata Hispanica Spanish Muske Rose. ——— } **R. moschata,** *Ait.*

R. holosericea Veluet Rose. 1085, 3. **R. muscosa,** *Ait.*

R. lutea Yellow Rose. 1085, 4. **R. lutea,** *Ait.*

R. pomifera The Pimpernell Rose. 1088, 3. **R. spinosissima,** *L.*

R. canina The common Sweete brier. 1087, 1. **R. rubiginosa,** *L.*

R. canina multiplex odorata The double Sweete brier. 1087, 1, *par.* 2. **R. rubiginosa,** *L. var.*

R. cinnamomea The cinnamon Rose. 1086, 5, *par.* 2, *descr. only.* **R. cinnamonea,** *L.*

R. cinnamomea flore multiplici The double cinnamon Rose. 1086, 5, *fig. descr. par.* 4. A variety of **R. cinnamomea,** *L.*

Rosmarinum Rosemarie. 1109, 1. **Rosmarinus officinalis,** *L.*

R. cachriferum Herbe Franckincense. 858, 4. **Cachrys Libanotis,** *L.*

Rubus Idæus The Raspis bush, or Hinde berrie. 1089, 2. **Rubus Idæus,** *L.*

R. saxatilis Stone Blackberrie tree. 1090, 3. **R. saxatilis,** *L.*

Rubia satiua Red Madder. 961, 1. **Rubia tinctorum,** *L.*

R. syluestris Wilde Madder. 961, 2. **R. peregrina,** *L.*

R. aquatica Water Madder. 961, 3? **Crucianella maritima,** *L.*

Ruta satiua Garden Rue. 1070, 1. **Ruta graveolens,** *L.*

R. syluestris Wilde Rue. 1070, 2. **R. montana,** *L.*

R. aquatica Water Rue. 1067, 1. **Thalictrum flavum,** *L.*

R. muraria Wall Rue, or Rue Maiden haire. 983, 3. **Asplenium Ruta-muraria,** *L.*

S.

Sabdariffa Thornie Mallow. 791, 2. **Hibiscus Sabdariffa,** *L.*

" . . . I had with great industrie nourished vp some plants from the seede, and kept them vnto the middest of Maie ; notwithstanding one colde night chauncing among many, hath destroied them all." *Ger.* 792.

Sabina vulgaris Common Sauin. 1193, 1. **Juniperus Sabina,** *β L.*

S. baccifera Sauin bearing berries. 1193, 2. **J. Sabina,** *L.*

Salix Rosea Rose Willow. 1204. A monstrous variety of **Salix alba,** *L.* ?

Salicornia Frog grasse, or Salt woort. 429, 1. **Salicornia herbacea,** *L.*

Saginæ Spergula Franke Spurrie. ——— *Ger. em.* 1125, 3. **Spergula arvensis,** *L.*

Saluia flore albo pinnata Winged white Sage. 623, 1? **Salvia officinalis,** *L.*

S. baccifera Sage bearing berries. ——— *Ger. em.* 765, 8. **S. triloba,** *L.*

S. maculata Spotted Sage. ———
S. crustatis oris Curled Sage. ——— } Varieties of **S officinalis,** *L.* ?

S. Italica flore candido aromatico, partim folio vulgaris. Italian white Sage. 624, *par.* 3 *or* 4.
S. grandiflora, *Ettling.*

S. Indica flore albo Indian white Sage. 623, 3. **S. officinalis**, *L. var.*

S. minor partim pinnata Pig Sage. 623, 2. **S. officinalis**, *L.*

Sambucus montana racemosa Mountaine Elder tree. 1234, 3. **Sambucus racemosa**, *L.*

S. rosea The Elder Rose, not rightly called the Gelder Rose. 1236, 2. **Viburnum Opulus**, *L. var.*

S. aquatica Water Elder. 1236, 1. **V. Opulus**, *L.*

19. *S. laciniatis folijs* Iagged Elder. 1234, 2. **Sambucus nigra**, *L. var.*

Sandalida cretica Square Codded Pease. ——— *Ger. em.* 1198, 3. **Tetragonolobus purpureus**, *Moench.*

Sanguisorba Great Burnet. 889, 1. **Sanguisorba officinalis**, *L.*

Sanicula vulgaris Sanicle. 801. **Sanicula europæa**, *L.*

S. guttata Spotted Sanicle. 644, 1. **Saxifraga Geum**, *L.*

Saponaria Sope woort. 360. **Saponaria officinalis**, *L.*

Satyrium odoratum Sweete smelling Satyrion. 172, 6. **Gymnadenia conopsea**, *R.Br.*

Satureia vera Winter Sauorie. 461, 1. **Satureia montana**, *L.*

Saxifraga alba White Saxifrage. 693, 1. **Saxifraga granulata**, *L.*

S. aurea Golden Saxifrage. 693, 2. **Chrysosplenium oppositifolium**, *L.*

S. Anglica English Saxifrage. 891, 1. *Ger. em.* 1021, 3, & 1047, 1. **Silaus pratensis**, *Besser.*

Scabiosa peregrina A strange kind of Scabious 585, 10. **Scabiosa cretica**, *L.*

S. maior Hispanica Spanish Scabious. 585, 9. **S. stellata**, *L.*

S. flore rubro Scabious with a red flower. 583, 6. **Trichera sylvatica**, *Schrad.*

S. marina Scabious of the sea. ——— **Scabiosa maritima**, *L.*

Scammonium Monspeliensium Scammonie of Montpellier. 716, 2, & 718, 3. *Ger. em.* 867, 3.
Cynanchum monspeliacum, *L.*

"2. . . . I haue plentie in my garden." *Ger.* 717.

"3. *Scamonea Valentina* Scammonie of Valentia." **Cynanchum monspeliacum**, *L. var.*

" . . . it is likewise found in the Iland of Candia, . . from whence I had some seedes, of which seed I receiued two plants that prospered exceeding well, the one whereof I bestowed vpon a learned Apothecarie of Colchester, which continueth to this daie, bearing both flowers and ripe seede. But an ignorant weeder of my garden plucked mine vp, and cast it away in my absence, in steede of a weede. . . . It flowred in my garden about Saint Iames tide, as I remember; for when I went to Bristow faire, I left it in flower; but at my returne it was destroied as aforesaid." *Ger.* 718. Johnson points out Gerard's transposition of the figures.

S. Syriacum verum Assyrian Scammonie. 716, 1. **Convolvulus Scammonia**, *L.*

Schœnoprasson Siues or Ciues. 139, 1. **Allium Schœnoprasum**, *L.*

Scordium Garlicke. 534, 1 & 2. **Teucrium Scordium**, *L.*

Scordothlaspi Garlicke Mustard. 204, 1. **Lepidium campestre**, *L.*

Scorodoprasson Garlicke Leeke. 139, 2. **Allium Ampeloprasum**, *L.*

Scorzonera Vipers grasse. 597, 1. **Scorzonera hispanica**, *L.*

Scrophularia Figwoort. 579. **Scrophularia nodosa**, *L.*

S. Indica Indian Fig woort. 579, *last par.* **S. lucida**, *L.*

Scorpioides Dodonæi Scorpion grasse. 267. ——— } **Scorpiurus sulcatus**, *L.*
S. bupleurifolio Broad leafed Scorpion grasse. ——— }

"I haue receiued seedes thereof from beyond the seas, and haue dispersed them through England." *Ger. l. c.*

S. Matthioli Scorpion grasse of Matthiolus description. ——— *Ger. em.* 338, 2. **Arthrolobium scorpioides,** *DC.*

Scorpioides Caterpiller Scorpion grasse. 266, *par.* 2, *no fig.* **Ornithopus compressus,** *L.*

Securidaca Hatchet Vetch. 1056, 1. **Securigera Coronilla,** *DC.*

Sedum maius Great Housleeke. 411. **Sempervivum tectorum,** *L.*

Seriphium Sea Wormwood. 940, 1. **Artemisia maritima,** *L.*

Serpentina maior Dragons. 682, 1. **Arum Dracunculus,** *L.*

Serpillum Wilde Time. 455, 1. **Thymus Serpyllum,** *L.*

S. citratum Wilde Time smelling like a Pome citron. ——— **T. citriodorus,** *Schreb.*

S. Pannonicum Wilde Time of Hungarie. 456, 3. **T. Serpyllum,** *β*, *L.*

Serratula Saw woort. 576, 1. }
S. flore albo White Saw woort. 576, 2. } **Serratula tinctoria,** *L.*

Sesamoides magnum Great barren Woade. 397, 3. **Thymelæa Tartonraira,** *All.*

S. paruum Little barren Woade, or Welde. 397, 4. **Catananche cærulea,** *L.*

" I haue had the seedes sent me from Padua in Italie. The flowers I do expect this present yeere." *Ger.* 397.

Seseli Aethiopicum frutex Shrub Hart woort. 1233. **Bupleurum fruticosum,** *L.*

S. Creticum Candie Hart woort. 894, *last par.* *Ger. em.* 1050, 1. **Tordylium officinale.** *B.*

S. pratense Field Hart woort. 891, 1. *Ger. em.* 1051. **Silaus pratensis,** *Besser.*

S. Peloponense Hart woort of Peloponesus. 893. **Thapsia villosa,** *L.*, or 903, 2. *Ger. em.* 1062, 2. **Melopospermum cicutarium,** *DC.*

Sisarum Skirrets. 871. **Sium Sisarum,** *L.*

Sison Wood Nep. 864, 1. *descr. not the fig.* *Ger. em.* 1016, 1. **Sison Amomum,** *L.*

Sida marina Marsh Mallow. 787, 1. **Althæa officinalis,** *L.*

Siciliana Parke leaues. 435. **Hypericum Androsæmum,** *L.*

Smyrnium Cræticum Alisanders of Candie. 869. **Smyrnium rotundifolium,** *Mill.*

Soldanella Sea Bindeweede. 690. **Convolvulus Soldanella,** *L.*

Solidago sarracenica Sarracens Consound. 347.

This name applies to **Senecio saracenicus,** *L.*, but Johnson points out Gerard's mistaken ideas of the plant, as shown in the description of *Epimedium* (v. p. 33 of the present work). He further states that he was credibly informed, that Gerard kept **Sisymbrium strictissimum,** *L.*, in his garden, as the true plant, *Ger. em.* 275 & 428, and probably sent it to Camerarius under that name, *v. Camer. Hort. Med.* 19.

Solanum hortense Garden Nightshade. 268, 1. **Solanum nigrum,** *L.*

S. somniferum Sleeping Nightshade. 268, 2, *not the fig.* *Ger. em.* 339, 2. **Withania somnifera,** *Dun.*

S. læthale Deadly Nightshade. 269. **Atropa Belladonna,** *L.*

20. *Solani somniferi similis fruticosa ignota planta, semine Constantinopolitano oriunda & delata a nobiliss. viro domino barone Eduardo Zouche, folijs tamen rotundioribus, & aliquantulum cauis* Shrubbie Nightshade. 277, 2, *fig. absent; given in Ger. em.* 348, 2. **Datura Stramonium,** *L.*

Sorbus torminalis The Seruice tree. 1287, 2. **Pyrus torminalis,** *Ehrh.*

S. Alpina Sapberrie tree. 1146, 2? **P. Aria,** *Ehrh.*

S. syluestris The Quicken tree. 1290. **P. Aucuparia,** *Gaert.*

Sophia chirurgorum Flixe weede. 910. 1. **Sisymbrium Sophia,** *L.*

Sorghum Turkie Millet. 77. **Sorghum vulgare,** *Pers.*

Speculum Veneris Venus Looking glasse. 356. **Specularia hybrida,** *A. DC.* ?
" I found it in a field among the corne by Greenehithe, as I went from thence toward Dartford in Kent, & in many other places thereabout, but not elsewhere: from whence I brought of the seedes for my garden, where they come vp of themselues from yeere to yeere." *Ger. l. c.*

Spondylium Cow Parsnep. 855, *descr. not the fig.* **Heracleum Sphondylium,** *L.*

Staphis agria Staues-aker. 398. **Delphinium Staphisagria,** *L.*

Stachis odorata Sweete smelling wilde Horehound. 563, 1. **Sideritis syriaca.** *L.*

S. Monspeliensis French Horehound. ——— *Ger. em.* 701, 2. **Phlomis Herba-venti,** *L.*

Stœbe Salamantica Great Siluer Knapweede. 590, 1. **Centaurea leucolepis,** *DC.*

S. argentea Little Siluer Knapweede. 590, 1. **C. alba,** *L.*

Stœchas Arabica Arabian Sticadoue. 469, 1. **Lavandula Stœchas,** *L.*

S. nudis cauliculis Naked Sticadoue. ——— *Ger. em.* 586, 4. **L. Spica,** *DC.*

Staphylinus Crœtica Carrots of Candie. 874. **Athamanta cretensis,** *L.*

Stramonium peregrinum Smooth Thorne Apples. 277, 1. **Datura Metel,** *L.*
" I have receiued seedes thereof from *Iohn Robin* of Paris, an excellent Herbarist, which did growe and beare flowers, but perished before the fruit came to ripenesse." *Ger.* 278.

Styrax The Storax tree. 1342. **Liquidambar styraciflua,** *L.*
" I haue two small trees in my garden, the which I haue recouered of seede." *Ger. l. c.*

Superba Austriaca A kind of iagged Pinke. 474, 4. **Dianthus superbus,** *L.*

S. pratensis Wilde field Pinks, or Cuckow flower. 480, 1. **Lychnis Flos-cuculi,** *L.*

S. duplici flore Cuckow flower double. 480, *bottom line.* **L. Flos-cuculi,** *L. var.*

Symphitum magnum Great Comfrey. 660. **Symphytum officinale,** *L.*

S. tuberosum Knobbie Comfrey. 661, *top par.* **S. tuberosum,** *L.*

S. petrœum Stone Comfrey. 507, 2. **Prunella laciniata,** *L.*

Syringa Italica White Pipe. 1213, 1. **Philadelphus coronarius,** *L.*

T.

Tabaco Indian Tabacco, or Henbane of Peru. 285. **Nicotiana Tabacum,** *L.*

Tamariscus Germanicus aut Narbonensis Germaine Tamariske. 1194. 2. **Myricaria germanica,** *Desv.*
" Fallitur." *Lob. MS.*

T. Italicus Italian Tamariske. 1194, 1. **Tamarix gallica,** *L.*

Tanacetum crispum Crispe, or curled Tansie. 525, 2. **Tanacetum vulgare,** *L.*

T. inodorum Vnsauorie Tansie. 525, 3. **Pyrethrum corymbosum,** *Willd.*

Tapsus barbatus Mulleine. 629, 1. **Verbascum Thapsus,** *L.*

Taxus Yew, or Yeugh tree. 1187. **Taxus baccata,** *L.*

Telephium semper virens Neuer dying Orpin. 417, 2. **Sedum Anacampseros,** *L.*

T. magnum Hispanicum Great Spanish Orpin. 416. 1. **S. Telephium,** *L.*

Testiculus odoratus Ladies traces. 167, 1, *left hand portion* ? **Herminium Monorchis,** *R. Br.*

Teucrium Tree Germander. 532, 2. **Veronica Teucrium,** *L.*
" I haue receiued of Master *Garret* Apothecarie one plant for my garden." *Ger. em.* 533.

Thalictrum magnum English great Rubarbe. 1067, 1. **Thalictrum flavum,** *L.*

T. paruum Little Rubarbe. 1067, 2. **T. minus,** *L.*

Thlaspi Candiœ Candie Mustard. 207. **Iberis umbellata,** *L.*

T. minus Little Mustard. 204, 4. **Lepidium ruderale,** *L.*

T. clypeatum Buckler Mustard. 205, 7. Clypeola Jonthlaspi, *L.*

T. vmbellatum Pesants Mustard. 205, 6. Iberis amara, *L.*

Thymum legitimum Time. 458, 2. } Thymus vulgaris, *L.*

T. durius Hard Time or Mother Time. 458, 1. "Fallitur." *Lob. MS.* } Thymus vulgaris, *L.*

T. durius alterum suavissimum Time of Candie. 459, 3. T. capitatus, *Link.*

Thapsia Stinking Carrots. 875, 1. Thapsia villosa, *L.*

Tithymalius paralius Sea Spurge. 401, 1. Euphorbia Paralias, *L.*

21. *T. characias* Wood Spurge. 403, 9. E. amygdaloides, *L.*

T. myrtifolius Mirtle Spurge. 402, 3. E. Myrsinites, *L.*

T. cyparissius Cypresse Spurge. 402, 5. E. Cyparissias, *L.*

T. dendroides Tree Spurge. 403, 7. E. dendroides, *L.*

T. tuberosus Knobbie Spurge. 407, 18. E. Apios, *L.*

Tilia The Line, or Linden Tree. 1298, 1. Tilia intermedia, *DC.*

Tormentilla Setfoile. 840. Potentilla Tormentilla, *L.*

Tordilion Bastard Cow Parsnep. 894, *par.* 1. Tordylium maximum, *L.*

Trachelium magnum Great Throate woort. 364, 1. Campanula Trachelium, *L.*

T. minus Small Throate woort. 364, 4. C. glomerata, *L.*

T. Giganteum Giants Throate woort. 365, 5. *Ger. em.* 448, 3. C. latifolia, *L.*

Tragopogon luteum Goats beard, or Go to bed at noone. 595, 2. Tragopogon pratensis, *L.*

T. purpureum Purple Goates beard. 595, 1. T. porrifolius, *L.*

Tragos Sea Grape. 959, 4, *no fig.* Salsola Kali, *L.*

Tragoriganum Goates Organie. 543, 1. Satureia Thymbra, *L.*

Tragium Germanicum Stinking Mother woort. 258. Chenopodium Vulvaria, *L.*

T. Bellonij Indian Mother woort. ——— Hypericum hircinum, *L.*

Tribulus terrestris Earth Calthrops. 1066. Tribulus terrestris, *L.*

"I found it growing in a moist medow adioining to the woode or Parke of Sir *Fraunces Carewe*, neere Croidon, not farre from London, and not else where; from whence I brought plants for my garden." *Ger. l. c.*

Trifolium bituminosum Treacle Clauer. 1019. Psoralea bituminosa, *L.*

T. fuscum Fower leafed grasse. 1028, 2. Trifolium repens, *L. var.*

T. Bristoliense Bristoll Three leafed grasse. ——— Possibly T. maritimum, *Huds.*, *v. Ger. em.* 1208, 6.

Triorchis lutea Yellow Ladie traces. 167, 2. Spiranthes autumnalis, *Rich.*

Tripolium magnum Great sea Star woort. 333. } Aster Tripolium, *L.*

T. paruum Lesser sea Star woort. 333, 2, *no fig.* } Aster Tripolium, *L.*

Tulipæ infinitæ Dalmatian cap, or Tulipa, in number and variable colours infinite. 116, *etc.*
Tulipa Gesneriana, *L.*, and T. suaveolens, *Roth.* ?

"*James Garret*, a curious searcher of Simples, and learned Apothecarie in London," had cultivated the various varieties of Tulip for more than twenty years. *v. Ger.* 117.

V.

Vaccaria Cow Basil. 395, 1. Saponaria Vaccaria, *L.*

Vaccinia nigra Hurtle berries blacke. 1229, 1. Vaccinium Myrtillus, *L.*

V. alba White Hurtle berries. 1230, 3. } V. Vitis-Idæa, *L.*
V. rubra Red Hurtle berries. 1229, 2.
Valeriana maior Great Valerian. 917, 1. Valeriana Phu, *L.*
V. Græca Greeke Valerian. 918, 5. Polemonium cæruleum, *L.*
V. Indica Indian Valerian. ——— Fedia Cornucopiæ, *DC.*
V. rubra Red Valerian. 550, 1. Centranthus ruber, *DC.*
V. aquatica Water Valerian. 917, 2. Valeriana officinalis, *L.*
Verbascum Matthioli French Sage. 625. Phlomis fruticosa, *L.*
V. Matthioli odoratum Sweete French Sage. 625. *par.* 2. P. fruticosa, *L. var.*
V. creticum Candie Mulleine ——— *Ger. em.* 459, 3. Verbascum spinosum, *L.*
V. nigrum Blacke Mulleine. 631, 2. V. nigrum, *L.*
V. fœmina Female Mulleine. 632, 4? } V. Lychnitis, *L.*
V. album White Mulleine. 632, 3.
Veronica mas Male Fluellin. 502, 2. Veronica serpyllifolia, *L.*
V. recta Pannonica Vpright Fluellin. 503, 5. V. spuria, *L.*
V. fœmina Female Fluellin. 501, 1. Linaria spuria, *L.*
Vinca peruinca Peruinkle. 747. } Vinca minor, *L.*
V. peruinca flore albo White Peruinkle. 747, *par.* 8.
V. peruinca flore purpureo Purple Peruinkle. 747, *par.* 9.
Violæ Marianæ variæ Diuers sorts of Marian Violets, or Couentrie bels. 362, 1 & 2. Campanula medium, *L.*
Viola calathiana Calathian Violet. 365, 6. C. glomerata, *L.*
Cf. Johnson, in *Ger. em.* 438, where he gives his reasons for doubting Gerard's knowledge of these plants.
V. lunaris perenius Neuer dying white Sattin. 378, 2. Lunaria rediviva, *L.*
22. *V. Theophrasti* Sommer fooles. 121, 3. Leucojum æstivum, *L.*
V. Hispanica Spanish Violets. 1043, 2. Lupinus luteus, *L.*
" Some haue called the yellow Lupine *Spanish Violets.*" *Coles, Adam in Eden*, 333. Gerard's name is not used by any other writer than Coles, so far as I can learn.
V. Matronalis varia Dames Violets, or Queenes Gilloflowers, diuers. 376, 1 & 2. Hesperis matronalis, *L.*
Violæ martiæ variæ March Violets, diuers sorts. 698, *etc.* Viola odorata, *L.*
Viurna Travellers Ioy. 739, 1. Clematis Vitalba, *L.*
Virga aurea Golden rod. 348, 1. Solidago Virgaurea, *L.*
Vitex Chaste tree. 1201. Vitex Agnus-castus, *L.*
Vites viniferæ variæ Diuers sorts of Vines. 724, *etc.* Vitis vinifera, *L.*
Vitis alba White Brionie. 720. Bryonia dioica, *L.*
V. nigra Blacke Brionie. 721, 1. Tamus communis, *L.*
Vrtica Romana Romaine Nettles. 570, 1. Urtica pilulifera, *L.*
Vua crispa baccis rubris Red Gooseberries. 1143, *par.* 7. } Ribes Grossularia, *L.*
V. crispa varia Diuers sorts of Gooseberries. 1143.
V. Zibeba The Vine that beareth corans of the shops. 726, 4. Vitis vinifera, *var.* apyrena, *L.*

X.

Xyris Stinking Gladen. 53. **Iris fœtidissima,** *L.*
Xanthium The Clot Burre. 664, 2. **Xanthium Strumarium,** *L.*
Xylon Bombast or Cotton tree. 753. **Gossypium herbaceum,** *L.*

" It groweth about Tripolis, and Alepo in Syria, from whence the Factor of a worshipfull merchant in London, Master *Nicholas Lete* before remembred, did send vnto his said Master diuers pounds weight of the seede, whereof some were committed to the earth at the impression hereof: the successe we leaue to the Lord. Notwithstanding my selfe three yeeres past did sowe of the seedes, which did grow very frankly, but perished before it came to perfection, by reason of the colde frostes that ouertooke it in the time of flowring." *Ger. l. c.*

Z.

Zyziphus The Beade tree. 1306, 1. **Melia Azedarach,** *L.*

Herbas, stirpes, frutices & arbusculas hoc Catalogo recensitas, quamplurimas ac fere omnes me vidisse Londini in horto IOHANNIS GERARDI chirurgi & botanici per-optimi, (non enim omnes eodem sed varijs temporibus anni pullulascunt, enascuntur aut florent) attestor

Matthias de Lobel.

Ipsis Calendis Julij. M.D. XCIX.

Across this certificate a pen has been struck, and " hæc esse falsissima " [attestor] " Matthias de Lobel," written at the end, in all likelihood by Lobel himself.

FINIS.

INDEX

TO THE MODERN NAMES CONTAINED IN THIS WORK.

ERRATA,

Which the Reader is requested to correct.

Page 26, line 29, for *Belgiœ* read *Belgiæ*.
,, 26, line 41, for *D. C.* read *DC.*
,, 27, bottom line, delete the comma after Chenopodium.
,, 31, line 18, for Colcichum read Colchicum.
,, 37, line 27, for comosus, read comosum.
,, 41, line 25, for Amydalus Persica read Amygdalus persica.
,, 43, lines 7 and 8, transpose M. and Muscari.
,, 49, line 16, for cinnamonea read cinnamomea.
On pages 23, 26, 40, *Mœnch*, should be *Moench*.